SOCIAL PROCESSES AND MENTAL ABILITIES IN NON-HUMAN PRIMATES

Evidences from Longitudinal Field Studies

Adult male of Arashima "A" at LaMoca (Encinal) site. Adult male is Japan born, resting in half-erect sit posture on a dead mesquite tree. Two and a half year old male companion chewing material from his cheek pouch, his fur is still wet from swimming.

Adult male of Arashima "A" troop in ball posture, engaging in self grooming.

Adult male of "A" troop displaying sentinel behaviour in a mesquite tree.

Young Adult Male "Ben" Macaca sylvanus

Young Adult Female Kin Group 'W.' Macaca sylvanus

SOCIAL PROCESSES AND MENTAL ABILITIES IN NON-HUMAN PRIMATES

Evidences from Longitudinal Field Studies

Edited by

Frances D. Burton

The Edwin Mellen Press
Lewiston/Queenston/Lampeter

Library of Congress Cataloging-in-Publication Data

Social processes and mental abilities in non-human primates : evidences from longitudinal field studies / edited by Frances D. Burton.

p. cm.

Based on papers presented at the Kempenfeldt meetings of the Canadian Association of Physical Anthropology in 1988.

Includes bibliographical references and index.

ISBN 0-7734-9537-1

1. Primates--Behavior--Longitudinal studies--Congresses.
2. Social behavior in animals--Longitudinal studies--Congresses.
3. Behavior evolution--Longitudinal studies--Congresses.
I. Burton, Frances D., 1939- .

QL737.P9S642 1992

599.8'0451--dc20 92-11041

CIP

A CIP catalog record for this book is available from the British Library.

The Edwin Mellen Press
Box 450
Lewiston, New York
USA 14092

The Edwin Mellen Press
Box 67
Queenston, Ontario
CANADA L0S 1L0

The Edwin Mellen Press, Ltd.
Lampeter, Dyfed, Wales
UNITED KINGDOM SA48 7DY

Printed in the United States of America

DEDICATION

With all my heart to my family:
Peter, Alexis and Leah
who goaded, discussed, encouraged, edited, and reassured.

TABLE OF CONTENTS

ACKNOWLEDGEMENTS

All but two of the chapters in this book were initially presented at the Kempenfeldt meetings of the Canadian Association of Physical Anthropology in 1988. I wish to acknowledge Prof. Larry Sawchuk for having inspired the symposium at those meetings which culminated in this book. A particularly exuberant thanks is extended to Alexis Silverman, for her volunteer labour in editing, for her competence and unflagging determination in what is generally, a grisly job.

Frances Burton

PREFACE

One area of primatology that has blossomed extraordinarily in the past ten years is the study of communication, cognition and social intelligence. We have thought for some time that primates are highly clever animals, because they do so well on our laboratory tests of intelligence. However, more recently, field studies have demonstrated that these animals also consistently employ higher-level cognitive skills in natural social situations. This volume, *Social Processes and Mental Abilities in Non-Human Primates*, now joins the ranks of a small but growing body of literature in primatology that celebrates the social and intellectual sophistication of our study subjects. The editor, Frances Burton, and the authors of the individual chapters of the book, are at pains to develop models of cognition that do justice to the complexity of the social context, the social knowledge and the social skills constantly at play in primate lives. Such an endeavor takes us far beyond the simplified, deterministic understanding of primate behavior that was prevalent earlier in our discipline's history.

This book also is an expression of a second significant trend in primatology, which is a growing attempt to revitalize research based on questions and theories to proximate explanation. The past twenty years have seen a remarkable development of an evolutionary perspective in primate studies, and of thinking based on ultimate, selectionist models. While this perspective has been productive and has led to many significant insights, at times the enthusiasm for adaptive, evolutionary explanations of behavior has threatened to drown out the voices of those who are still interested in the here and now of primate lives. Thus, books such as this are an important reassertion of the value of understanding the immediate processes (e.g. cognitive, perceptual, environmental, experiential, motivational, and social processes) at work in the lives of non-human primates. At the risk of oversimplifying, it is worth noting that the authors of this volume are primarily anthropologists and psychologists. It is likely not a coincidence that whereas zoologists have been influential in promoting evolutionary models in primatology, anthropology and psychology are disciplines that have continued to contribute ideas and theories about proximate causation and the researchers interested in pursuing these ideas. Surely, primatology can benefit from the confluence of these different perspectives.

Finally, and not least, this book is an expression of an entirely new phenomenon in primatology, and that is a compilation of works authored entirely by Canadian primatologists. When I first came to Canada in the early 1970's, I was already familiar with the research by Frances Burton on Barbary macaques. Over the succeeding years, mainly at the annual meetings of the Canadian Association for Physical Anthropology, it has been my pleasure and my gain to meet the editor and her students, now employed at universities across the country. Primatology is alive and well and living in Canada, It is my hope that the present volume will help to bring this fact to the reader's attention, in addition to achieving the author's explicit goal of conveying the "richness and texture" of primate social cognition to the reader.

Linda Fedigan
Edmonton
September 1991

INTRODUCTION

by

Frances D. Burton

Most of the chapters of this book were originally presented at a symposium of the Canadian Association of Physical Anthropology in 1987. The interest was in exploring the meaning of cognition, of mental processing in non-human primates. Because simple deterministic models did not seem to adequately describe what the authors had observed in their various research areas, the object was to present a forum where the ideas developed during research could be brought to bear on this issue. The theme which all the authors addressed is that non-human primates act on the basis of mental evaluations; that they are cognitive organisms capable of "higher order" mental processes. Interest in this issue goes back to the earliest researches on non-human primates, when psychology was investigating the nature of non-human primate intelligence, its differences and relationship to human intelligence. Over the four decades that primates have been seriously studied in the field, there has been a growing certainty that mental processes as humans experience them are something we share with non-human primates from our primate heritage. Within the past half-decade, a number of books have been published that deal with cognition, or mentality in monkeys and apes. The contribution here adds to that store with a set of chapters that explore cognition from different perspectives. Most of the authors are anthropologists, and raised with an appreciation for the concept of 'culture' as patterns and norms of group behaviour. Their interests in cognitive processes share several premises and explore similar questions: [1] that a society is a set of coordinated individuals; [2] that the focus of study is the individual and the interactions between individuals that compose the society; [3] that these interactions constitute unique, local patterns of doing things; and [4] that the implication of being a cognitive being emphasizes the impact of group local behaviour patterns on the social development of the individual.

The authors present varying views, and support different paradigms about behaviour. Chan's chapter establishes the argument that what non-human primates are doing in their groups is more than merely finding ways of eking out an existence. He notes that social relations go beyond the consequence of ecological pressures; that what goes on inside the group, the result of interactions between individuals, is a result of cognitive recognition of each other. By their very brain structure and function, primates are unique amongst mammals for their cognitive capacity. Amongst their capabilities is the ability to draw inferences concerning social relations, to predict outcomes, and to interpose information to ensure the direction of the outcome. Chan gives a case in point when, in reference to primate vulnerability to predators, he asks: "Could the fear of being excluded from the group be a means of social control exercised by the group? In this case predator pressure can not be measured without reference to the process of social interaction." Clearly, what happens in the social group, its history of affiliations, friendships, political maneuverings, or antagonisms, will bear on who gets preyed upon.

Burton details how the group is social and the cognitive consequences of sociality. I discuss the social group as an integrated network, not a set of individuals facing individual problems and creating individual solutions. The consequences of interactions are standards of behaviour, which, as they become conventionalized within a group, becomes its tradition, differentiating it from any other group. The social intelligence displayed by non-human primates goes beyond recognizing individuals, to recognizing their representations, that is the manufacture of images constructed by the individual. Amongst primates, individuals are more than 'male,' 'female,' 'pregnant,' 'infant,' etc., individuals are recognized as the set of behaviours associated with them. The value of a piece of information is associated with the holder of that information. Following 'leaders' to food is that sort of process. Non-human primate communication systems are multi-faceted. The various channels or sensory pathways permit information to be coded in what are termed modalities.

They permit, as Zeller outlines in her chapter, transmission not only of feeling states and subjective representations, but also of information external to the individual. Complex messages exist because disparate elements can be joined in unique, personal ways and these can *nonetheless* be interpreted accurately. Her approach to the study of communication includes analysis of the individual's construction of the message; "the level of volition and control over sending that individuals possess." Zeller shows how eye components of facial gestures can be mixed and matched with mouth components such that new statements appear, whose meaning, however, is clear. Meaning and interpretation are 'read' from the message by the receiver who evaluates not only who the sender is with reference to the sender's status, age, gender, previous behaviour and reliability but also in terms of the previous relations the receiver has had with the sender.

In addition, the encoding of the message, its channel, to whom it is directed, and its intensity, are all matters of choice on the part of the sender. The choices themselves are indicative of the range and complexity of the knowledge a primate member of society possesses. Acknowledging that animals wandering through forest or savanna require a great deal of environmental knowledge to live, is not a new idea. In the 1960's, at the very beginning of serious field research, Hall worked in the field with baboons and patas monkeys. He appreciated not only the quantity and complexity of knowledge required just to get on, but also that there were patterns to knowing itself. Hall labelled the norms of knowledge "conformities" and, most importantly, he recognized that these "conformities" extended to codes of social behaviour and social knowledge within the group.

In a paper published after his death in 1965, he included amongst social conformities: familiarization with others in the group, appreciation of social distance and social discriminations, i.e., interaction patterns (Hall, 1968). He was a pioneer in analyzing the *process* of social learning and discussed social facilitation, observational learning and modelling of behaviour as important aspects of this process. These insights are fundamental to current research. In their discussion of what it is that orangutans do, Galdikas and Vasey demonstrate the range of information an orangutan must have to make a living in the rain forest where resources are variably distributed. They utilize Menzel's (see references at the end of the book) rigorous criteria of what constitutes environmental knowledge and how we would recognize it. For example, they concur that taking a shortcut is the "clearest" evidence of the ability of non-human primates to map their environment because: "This behaviour entails the need for detailed individual memory of environmental features (criterion 1) and the ability to represent several such features in quantitative terms so that at least distance policy can be employed (criterion 1 & 4)." Indeed, "Orangutans seemed to be better at this than primatologists...." Primate intelligence may well have developed in the service of **prediction.** The ability of primates to store thousands of images is a quality of the visual centres in a brain which is different to that of other mammals. Primates are able to recall and associate these images quickly. By comparing images (visual, olfactory, etc.) with stored ones, primates can predict a situation. Fruiting of trees, for example, is predictable from variables encountered in the past (e.g., Burton, 1972; Richard, 1985; Milton, 1988). Orangutans, as Galdikas and Vasey show, have fine tuned this ability in a forest where resources are widely distributed.

As it concerns behaviour, the ability to store images enhances social knowledge. Individual members are classified (Burton, 1984) and known from physical attributes (Kummer et al., 1990). In addition, the ability to predict means applying learned, social rules to a new, but familiar situation. In addition, it permits the primate to arrange elements of behaviour, to ensure a desired outcome. This requires knowledge of one's self in relationship to others; the ability to predict the consequences of certain representations or communications. If the elements, for example, of a meal, the cutlery, dishware, serviettes, are arranged according to knowledge of what that arrangement means, they will evoke a predictable response in the guests (awe, pleasure, shock). The manipulation of the signs depends on mutually held signification: linen tablecloth versus plastic; silver cutlery vs.

stainless. Galdikas and Vasey detail how orangutans use social information to manipulate the behaviour of conspecifics as 'social tools' for personal benefit.

In describing the nature and extent of social intelligence in orangutans, they show how intimate knowledge of members of the social network and of their probable behaviour is required in these animals who live widely dispersed from one another. It is awareness of oneself, others, social interactions, and one's interactions with others that constitutes the particular intelligence of non-human primates. Paradoxically, this is the domain of *proximate behaviour* (Kummer et al., 1990). This is evident in orangutans, where "social behaviour has utility to others" because observation of the behaviour of others means that what is performed is retained and disseminated throughout the group. This provides a rapid means to adapt to local exigencies. Patterns "invented" locally by one individual can have far-reaching effects on the particular group. Such a form of adaptation has its advantages.

The development of intelligence has long been considered to be social (e.g., Kummer et al., 1990; Washburn and Hamburg, 1968). Recently that origin of primate intelligence has been given a more restricted basis. Whiten and Byrne argue "primate intelligence is not just 'social,' but Machiavellian in its origins" (Whiten and Byrne, 1988a: 6). They derive the term from Niccolo Machiavelli, who, in the 16th century, advised Cesare Borgia, his Prince, that generally considered immoral behaviour (e.g., murder) may be undertaken for the *State* if the end justifies it.

"Machiavelli" thus has long been synonymous with deception, hypocrisy and even evil. It subscribes to the popular view about "cleverness" which presumes that being clever means keeping some information away from others. The display of cleverness in non-human primates is explained in this popular view and incites the label long associated with deceit and self-interest to be attributed to them. Ascribing the development of intelligence to the service of personal gain identifies, however, only an aspect of social intelligence. Social intelligence, while less "catchy," is a term preferred to "Machiavellian Intelligence" as it is a neutral one which does not judge the motivation in assessing the act. More to the point, social intelligence need not be self-serving or deceptive (that is, concealing a truth or purposely misleading). While primates may deceive others about the location of, for example, ripe fruit, individuals will also signal others about the availability of ripe fruit at a particular location. The opinion formed about the *meaning* of calls that inform of location depends on the 'translation' given them by the observer: "I am happy that there is fruit" is qualitatively different to "Look everyone, there is fruit over there." Since "We do not sufficiently know what happened among our subjects last night and last year..." (Kummer et al. 1990: 92) caution must be exercised before interpretation. Without knowledge of the history of each individual, his/her group, and the history of relationships between individuals it extends credulity to attribute motivation.

The relationship of the acts, including communication, of one individual on the behaviour of a witness, is the subject of Chadwick-Jones' contingency model. In focusing on the dynamic between two individuals, he explores social exchange and reciprocity beyond kin relations. These depend on the awareness that individuals have of the personality of others as well as the history of the relationships between them. Chadwick-Jones proposes a method by which such interactions can be studied. His social contingency model follows a dyad, providing a continuous record of reciprocal actions. The sequence of action, reaction, describes a 'zig-zag' which can then be analyzed in detail. The model permits examination of the nature and extent of social exchange networks. Proximity, or grooming, for example, are social objects that can be exchanged. Exchange, Chadwick-Jones suggests, is "... a reciprocal transaction with the end result of gain, in some degree, for each of the parties." Primates can exchange or transact proximity, coalitions and alliances, or friendship. Such interactions constitute proximate behaviour. The proximate world is the here-and-now of primate behaviour. It is the locus and time of their activities. Ultimate behaviour is the label given to behaviours deemed by the observer to have evolutionary significance. Behaviours that relate to successful promulgation of the individual are considered ultimate behaviours. Clearly, there is overlap - and occasionally confusion - between the two as value or motivation is imputed by the observer to the animal. In addition, as Paterson suggests in his chapter: "The usual state of affairs is that these contributing factors account for only part of the total proximate causality...." Paterson explores this realm where behavioural modification is insufficient, so that a physiological change may accommodate the situation, and only if the selective pressure is sufficiently large, or of a form which bypasses these mechanisms, may it have an actual selective effect upon the structure of the population's gene pool. Paterson enters a caveat: that concern with cognitive processes, social intelligence and the like should not ignore internal adaptations, or physiology. As a behavioural ecologist, he reckons the individual responds to the environment physiologically, and that behaviour is the adaptive mode permitting such adaptation. In his 'shell' model of behaviour, Paterson uses the concept of deep and surface structures, where deep structures are those of evolutionary significance, and surface structures a local, particular expression. The construct derives from structural linguistics where not only the capacity but the structure for language lies within the human brain, and languages are the superficial variants derived from that structure.

The layers of Paterson's model are labelled as self-awareness, cognition, and group-specific learned behaviour. Field observers can only examine this outermost of protective shells, the organism's behavioural activity, and must infer the functioning of the various layers within. The consequence will be interpretation of the visible without appreciation of causality. Responses to biological "imperatives" (as they were once called), are behavioural and they in turn create their own reality to which subsequent behavioural responses accrue. These give rise to local patterns of behaviour, or traditions.

A central issue in the study of behaviour has been variation. What is the meaning and extent of species-specific behaviour? If behaviour is species specific (and therefore taxonomically valuable), variation would be minimal. However, field workers find important differences between local groups of the same species, sometimes not more distant than half a kilometer apart. What could account for this? Founders bringing their personal memories of group behaviour to a new locale would enact only that behaviour which they had remembered. Clearly, personal memory would be only a part of the total repertoire of the group. Hence, new behaviours would develop from amongst the inventory as presented by the founders. In addition, significant members of the group would display behaviours which would be prominent because that individual was highly visible. Those behaviours would influence and ultimately be stored by the young. This process is termed "Tradition Drift" (Burton, 1972).

Tradition Drift as a paradigm of behaviour competes with the dichotomization now current. "Proximate" and "ultimate" behaviours or "surface" and "deep" structures are attempts to establish causation on the evolutionary level of observed behaviour. Behaviour is no longer seen as meaningful to itself. The attribution of function to each feature of behaviour in the service of evolutionary design is reminiscent of theological arguments over free-will and ultimate destiny. These formulations subscribe to a view of behaviour as ordered because it is ordained. Variation, like chaos, is acceptable only if trivial, i.e., surface.

In her chapter, Hornshaw notes the need in contemporary behavioural studies to rid themselves of human cultural bias in order to really 'see' non-human primate behaviour in all its variations. Hornshaw studied wild and captive macaques and found that local traditions differentiate between groups but whereas the process of tradition formation is generally acceptable for 'wild' groups of the same species, there has been reluctance to accept the differences in captive groups as 'local traditions.' "It might be that in spite of shared species, shared management decisions and a shared social history, different groups, as a result of processes uniquely internal to each group, deploy themselves differently."

The expectation is that captive behaviour is homogeneous and qualitatively different to wild behaviour because managing the organisms "alters the parameters of the opportunities and constraints of the environment." Adaptation is thought, therefore, to be obviated. The problem, however, is not with the *process* but rather with unquestioned assumptions.

Hornshaw notes that since efficiency or economy is 'good,' nature and natural processes which are always deemed 'good,' will be efficient or economical. What she terms "formulaic primatology" is evidenced in the reduction of social behaviours to simple 'economic' responses and she cites the work on lactating baboon mothers as an example where calculation of caloric expenditure neglects the social, cognitive bases of this mother-infant interaction. Hornshaw affirms Kummer et al.'s (1990) instruction to recognize the proper use of terminology, cautious methodology and control. In addition, what is efficient is not necessarily economical: redundancy in communication, for example, is most efficient but not necessarily economical. Redundancy in biological systems, too, ensures delivery of a message, even though extra energy is thereby expended.

Mechanistic paradigms have tended to underestimate the richness and texture of social behaviour. Paradoxically, early workers, in their descriptions of primate societies, preserved precisely these qualities. These seven chapters constitute an addition to the growing interest in the area of primate studies which recognizes non-human primates as animals functioning on a complex level based on their cognitive abilities.

CHAPTER ONE

Problems with Socioecological Explanations of Primate Social Diversity

by

Leslie K. W. Chan

Introduction

Primate societies are highly complex and variable. Although this statement appears to be intuitively obvious, early primate researchers (e.g., Zuckerman, 1932; Carpenter, 1934, 1942, 1964; Washburn & Devore, 1963; Hall, 1965) hardly suspected the amazing array of social organizations that were to be discovered in the ensuing decades. As the scope of field research expanded and data accumulated at an accelerated pace, it became apparent that considerable variation in primate social organizations exist both within and between species (Rowell, 1969, 1972; Struhsaker, 1969; Eisenberg et al., 1972). The question as to why primates live in groups and the causes of social variability demand explanation. The search for rules governing primate social organization has become the major preoccupation of primate research to date (see the following edited volumes for examples: Clutton-Brock & Harvey, 1977; Sussman, 1979; Tuttle, 1975; and the reviews by Richard, 1981; Terborgh & Jason, 1986; Crook, 1989; Devore, 1990). Early primate studies were, however, largely descriptive and expressed little concern for evolutionary causes of social organization. The initial impetus for an evolutionary approach to primate social organization was provided by John Crook and Stephen Gartlan (1966). Modelling on Crook's earlier success with the explanation of weaverbird social variability (Crook, 1964), the authors argued that it is possible to predict various aspects of social organization on the basis of a small number of ecological variables, primarily food distribution and predator pressure. More specifically, Crook and Gartlan classified approximately thirty primate species into five adaptive "grades," representing an evolutionary sequence of simple to complex "socio-types."

Although Crook and Gartlan's synthesis was deficient in several aspects, such as a scanty data base and qualitatively broad ecological and behavioural categories that tend toward oversimplification, their work soon became the basis on which numerous field studies and theoretical discussions of primate social organization were based (e.g., Struhsaker, 1969; Crook, 1970; Eisenberg et al., 1972; Altmann and Altmann, 1970; Altmann, 1974; Aldrich-Blake, 1970; Kummer, 1971; Dunbar and Dunbar, 1974; Hladik, 1975; Clutton-Brock and Harvey, 1977; Wrangham, 1980). By the mid-1970's, there appeared to be sufficient consensus on the evolutionary significance of social organization, that E.O. Wilson (1975: 255) felt the following statement was warranted:

> The principal organizing concept in the study of primate societies has been the theory that social parameters are fixed in each species as an adaptation to the particular environment in which the species lived. The parameters include size, demographic structure, home range size and stability, and attention structure.

Not surprisingly, many of the works in primate socioecology during this period are concerned with finding correlations, though with varying emphasis, between group size, ranging pattern, diet, predator pressure and mating system. The implicit assumption is that once the correlations are found, the evolutionary causes of social variability can be identified. Relatively few efforts, however, have been made to explain *how* various social organizations actually arise as a consequence of species interaction with the relevant ecological variables (Richards, 1981). An important corollary to the view that social organizations of animals are evolutionary responses to a limited set of ecological conditions is the proposition that similar social structure will result when animals, *regardless of their phylogenetic background*, are faced with similar ecological situations. As Emlen (1980: 129) has explicitly stated:

> Animals faced with similar ecological "problems" exhibit a predictable convergence in their "solution" as shown in their social organizations.

This proposition is certainly suggestive and its wide acceptance has strongly encouraged the continued search for broad correlation between social organization and ecology across a wide range of primate species. The purpose of this paper is to evaluate the general propositions as stated by Emlen and Wilson, among others, in the context of two broad questions:

[1] Can ecological factors alone explain why primates live in complex social groups?

[2] In addition to ecological factors, what would be the necessary conditions for stable patterns of interaction to be maintained among the members of a social group?

I shall argue that while ecological variables such as predation pressure may account for aspects of social structure such as group size, it may not explain why members within a group maintain stable relations with each other. This points to the need for proper delineation of the level of organization at which explanation should be directed. A proper evaluation of primate social variability should be based on a clear understanding of the differential effects of the environment on the individuals, the breeding unit, and the social group, both separately and as a whole. The tendency to confuse the various levels of organization and the failure to consider the phylogenetic history of the species being compared may result in the generation of correlations where none actually exist. In many cases, the similarities between species, such as the presence of a single-male multi-female structure, may be due to common phylogenetic heritage (Struhsaker, 1975). While in other cases, the similarities may be more apparent than real considering other aspects of the social organization that may have profound effects on members of the different species being compared.

Classification of Social Organization

Despite the many theoretical attempts to correlate social organization with ecological variables, few authors actually bother to provide an operational definition of what constitutes social organization. This is due, in part, to the fact that the spatial relations of most open country monkeys are easy to recognize and thus definitions of social grouping and organization were thought to be superfluous (Struhsaker, 1969). Consequently, ideas concerning grouping and social organization remain rather vague and imprecise. The lack of a set of standardized definitions of social systems may well constitute the major weakness underlying studies of comparative primate ecology since correlation of social organization and ecology may be based on widely different criteria that may not be strictly comparable (Rowell, 1972). As summarized by Fedigan (1982: 44), primate social organization has traditionally been classified in terms of any one, or various combinations of, the following variables:

1. the group size: large troops, medium-sized groups, small units;
2. the group composition: multi-female, multi-male groups; uni-male, multi-female groups; uni-female, uni-male groups;
3. the mating system: multiple mating by both males and females, polygynous, monogamous;
4. the social role, especially for adult females and adult males, in relation to socialization, resource and predator defence, group movement and leadership, etc.;
5. the major internal structuring principles, be they various types of dominance hierarchies, kinship networks, age/sex class associations, or alliance system;
6. the permanence versus instability of group membership;
7. the tendency or ability of one level of social group to aggregate into larger social groupings;
8. the presence of only heterosexual reproductive units; or the additional presence of all-male groups; all-female groups; single individuals;
9. the pattern of interaction - who does what to whom, how often and under what circumstances?

The choice of variables is, to a large extent, dependent upon one's prior theoretical commitment and its underlying assumptions. Researchers who favour ecological parameters as prime determinants of social organization have tended to focus on mating systems as the defining criterion of social organization. This is because the mating system is thought to be highly correlated with group size and composition, which in turn are most directly influenced by ecological factors (Orians, 1969; Goss-Custard et al., 1972; Emlen & Oring, 1977; Wrangham, 1979, 1980).

Following this line of reasoning, Eisenberg and his colleagues (1972) developed the following schemes of the evolution of primate grouping patterns:

I. solitary, including mating adults as temporary social groups.

II. monogamous, adult pairs with offspring.

III. single male group with one breeding adult male and bonded females and offspring.

IV. age-graded, male group consisting of one adult male, his male offspring, bonded females and offspring.

V. multi-male group, consisting of several adult males hierarchically ranked and females with their progeny.

The trend in the evolution of primate social systems, Eisenberg et al., (1972) argued, lies in the increasing tolerance of adult males for stable social living, first with adult females and later with other adult males. Factors that affect the tolerance level of breeding males are therefore key factors that restrict or promote the formation of new social groupings.

In all models of primate social evolution, group living is taken to be the necessary first step for the development of more complex systems. It is therefore critical to ask why primates, and specifically all diurnal species, live in groups in the first place.

Why Primates Live in Groups

Many studies of primate socioecology have been inspired by the general recognition that there are major costs as well as benefits to living in groups (Crook, 1972; Crook et al., 1976; Schoener, 1971; Bertram, 1978). As discussed by Alexander (1974), the costs associated with group living are:

[1] increased conspicuousness to predators,

[2] increased competition within the group for food, mates or other limited resources,

[3] increased risk of infection by contagious diseases and parasites.

On the other hand, group living has several important benefits, the most crucial of which are:

[1] reduction in predator pressure resulting from increased vigilance and group defence against predators,

[2] improved foraging efficiency for clumped ephemeral food resources,

[3] improved defence of limited resources, including mate and food sources, against other groups of conspecifics.

Group living is generally expected when advantages obtained through sociality outweigh the potential liabilities (Schoener, 1971; Alexander, 1974; Bertram, 1978). Furthermore, the combination or balance of different kinds of costs and benefits will allow one to predict the optimal group size for a given environment as well as to explain why some animals defend resources in an exclusive range. In other words, assessment of the interplay between such factors as predator avoidance and distribution of key resources will provide a better understanding of the ecological causes of group size and cohesiveness.

In primate socioecology, two major alternative explanations for group living have been proposed. Following the suggestion of Alexander (1974), van Schaik (1983) and Terborgh (1983) have argued that predation pressure has been the single most important factor in promoting group living in primates, and the advantages accrued outweigh that of intra-group competition for food. On the other hand, Wrangham (1979, 1980, 1982) has argued that the role of predation pressure is negligible whereas cooperative defence of key resources is the main force promoting group living in primates. Many authors, it should be noted, acknowledge the importance of predation, while presuming advantages to feeding in a group as well (Eisenberg et al., 1972; Altmann, 1974; Clutton-Brock, 1974). Although each of these explanations differ in their emphasis of ecological variables, they share the similar assumptions that ecological factors play a deterministic role in shaping social systems and that social systems are products of the individual's interaction with the environment. Each of the hypotheses outlined above will be described and the strength and weakness of the underlying assumptions will be discussed.

Effects of Predation on Group Size and Composition

It is generally assumed that the specific effects of predation on group size and composition of different primate species will depend upon the form and intensity of predation (Devore and Hall, 1965; Crook and Gartlan, 1966; Eisenberg, et al., 1972; Alexander, 1974). Several predictions have been derived from this general proposition.

(1) Group size and stability: Group sizes should be large where risk of predation is high and small where risk is low. Predation pressure is thought to be highest for terrestrial species occupying open-country savanna habitats, terrestrial species are therefore expected to live in large groups. In their study of savanna baboons (*Papio cynocephalus*) at Amboseli, Altmann and Altmann (1970) report that 25% of all deaths during the 14-month study period were due to predation. They also noted that all female baboons and most males are found in spatially cohesive groups of about 20 to 90 animals. But groups as large as 200 and as small as 13 have also been seen.

The large group size of the savanna baboon as well as other terrestrial species such as the rhesus macaque (*Macaca mulatta*), the gelada baboon (*Theropithecus gelada*), and the patas monkey (*Erythrocebus patas*) has been considered a response to greater predator pressure in open country where refuges are rare. A specific advantage to these groupings is increased vigilance and therefore earlier detection and warning of potential predators (Crook and Gartlan, 1966; Eisenberg et al., 1972; Kummer, 1971; van Schaik, 1983). The specificity of alarm calls among vervet monkeys indicates the advantage accrued to terrestrial primates by alerting group members to the presence of predators (Struhsaker, 1967; Seyfarth et al., 1980).

A large group also decreases the risk that any one individual will be the one preyed upon by the predator (e.g., Terborgh, 1990; Clutton-Brock and Harvey, 1977; Crook et al., 1976). This suggestion is based in part on the "selfish herd" principle advanced by Hamilton (1971). Furthermore, group living also makes possible different forms of mobbing and group defence against would-be predators. Such mobbing behaviour and actual attacks of potential predators have been observed among baboons and chimpanzees (Kortlandt, 1967).

In contrast to terrestrial primates, predation on forest primates is much harder to detect and appears to be much more rare. It is therefore generally assumed that predator avoidance played at best a minor role in the evolution of social organization among forest-living primates (Crook and Gartlan, 1966; Crook, 1970; Aldrich-Blake, 1970; Struhsaker, 1969). Groups tend to be small among forest-living primates and a number of solitary as well as monogamous species (Eisenberg, 1978) can be found. In contrast, terrestrial primates tend to live in multi-male multi-female groups, with few or no solitaries (Altmann and Altmann, 1970; Fedigan, 1982).

(2) Group composition: It has been suggested that higher predation pressure should lead to a larger percentage of males within a group, assuming that males are the prime defenders of the group (Devore, 1963; Kummer, 1971; Eisenberg et al., 1972). Crook et al., (1976) suggest that the mere presence of a male may increase the safety of his females and young. But "whether or not this is likely will depend on basic characteristics of the animal in relation to the hunting characteristics of the predators" (Crook et al., 1976: 313). The manner in which primates respond to a potential predator also depends partly upon their body size and mobility. Thus patas monkeys, which are small and agile, respond primarily by flight and the males usually assume the role of lookout (Hall, 1965; Gartlan and Struhsaker, 1975); baboons, which possess larger body size and canines, may engage in group mobbing or outright attack of the predator (Altmann and Altmann, 1970; Hamilton et al., 1976).

(3) Sexual dimorphism: Animals who live under strong predation pressure should exhibit a strong degree of sexual dimorphism. Males of a species living under this condition should be larger (e.g., body size, dentition, temporal musculature) and better equipped to fight (Crook and Gartlan, 1966; Eisenberg et al., 1972; Clutton-Brock and Harvey, 1977). Devore (1963: 345) has stated that "the trend towards increased fighting ability in the male of terrestrial species is primarily an adaptation for defence of the group." The increase in sexual dimorphism is seen to be correlated with

dominance behaviour, which is considered to be a controlling mechanism in the social organization of the group. It is widely assumed that a social group in a harsh habitat will benefit from a tightly controlled organization where the "male is influential in promoting cohesion of the troop and serves as a focus for the troop's movement" (Eisenberg et al., 1972: 136). Predation pressure, therefore, not only effects group size but may have different influence on the internal organization of the group as well.

Although these predictions concerning the effects of predation are highly suggestive and appear plausible, none of them have actually been tested. In addition, there are major weaknesses associated with them:

(1) Lack of empirical data: Although Eisenberg et al., (1972: 137) considered the adaptive advantage of multi-male groups in terms of antipredator behaviour, they admitted that "no single aspect of primate field studies has less supportive data than the generalizations concerning the survival value of the various presumed antipredator mechanisms." The current situation does not differ a great deal, despite the large number of field studies being conducted (Richards, 1985; Box, 1984).

Primate researchers often report the presence of "potential" predators within the range of their study animals. However few actual attacks have been observed. For example, Dittus (1975, 1977) studied toque macaques (*Macaca sinica*) in Sri Lanka, where he noted the presence of jackals (*Canis aureus*), fishing cats (*Felis viverrina*) and the jungle cat (*Felis chars*). Although these animals are carnivores and are sufficiently large to be threats, Dittus encountered no interaction between them and the monkeys. Muckernhirn and Eisenberg (1973) recorded fifty-one cases of predation by the Ceylon leopards in an area where they were sympatric with rhesus macaques, but no predation on the latter was observed. In a recent attempt to assess the impact of predation on primate evolution, Anderson (1986) amassed a wide range of field reports that make reference to predators, both actual and assumed. Of particular interest is the large number of references to the domesticated dog as predator. Many macaque populations near human settlements are indeed vulnerable to free-ranging dogs. Such dogs may disrupt the daily routine of the monkeys and cause considerable distress to mothers with young infants. It is doubtful, however, that they prey on monkeys (Lindburg, 1977). It is important to note that non-human predators have been eliminated from many areas where primates have been studied. Consequently, the impact of non-human predators is slight in comparison to that of human hunting and systematic trapping. How primates have adapted to constant habitat destruction and human predation is an area that is seldom explored and clearly deserves further attention (Richard et al., 1989). These examples point to the difficulties of assessing "potential" predators, let alone defining and measuring predation pressure. Moreover, in view of the lack of quantitative data on predation rate, the assessment of the impact of predation on demographic parameters is at present not possible (Anderson, 1986).

(2) Problem with correlation: In the absence of actual predation rate, many have turned to group size, which is more amenable to quantification, as an indicator of predation pressure. As pointed out earlier, the generalization that larger group size is to be seen in terrestrially adapted groups has been favoured by many workers. This is based on the further assumption that predation risk is higher in terrestrial environments. Although comparative data appear to support this view, there is considerable variation within and between species that remains to be explained. In a survey of forest-dwelling cercopithecine species in West Africa, Struhsaker (1969) reported that mean group size was 10-12 for most arboreal groups studied. However, *C. l'hoesti* is the major exception with an average group size of 5-6 individuals. *C. l'hoesti* also differs from other cercopithecines in spending most of their time foraging as well as fleeing from danger on the ground.

Patas monkeys and baboons are both well-known savanna dwellers, but the mean group size of the former is almost half of the latter. On the other hand, olive baboons (*Papio anubis*) living mainly in the forest have mean group sizes equal to or exceeding those of their savannah-dwelling conspecifics (Dunbar, 1988). It is obvious that while group size may be affected by predation pressure, the simple correlation of group size with habitat types (e.g., terrestrial versus arboreal) is grossly inadequate. The lack of such correlation calls into question the assumption that group size is an adaptive response to predation pressure.

(3) Inadequate testing procedures: The idea that large groups increase vigilance and hence earlier detection of predators has been tested for some birds (e.g., Bertram, 1980; Barnard, 1980) and mammalian species (Hoogland, 1979; Clutton-Brock et al., 1982; Lipetz & Bekoff, 1982). Verification of this proposition is based on the further assumption that visual "scanning" constitutes vigilance with respect to predators such that the amount of time spent "scanning" should correlate with group size. Recently, De Ruiter (1986) has attempted to test this assumption in two groups of wedgecapped capuchin monkeys (*Cebus olivaceus*). He concluded that members of the "large" group (n+25) spent less time scanning while members of the "small" group (n=8) spent more time scanning in order to achieve the same level of vigilance as the large group. While there appears to be a correlation between group size and time spent scanning, it remains to be demonstrated that predation is the cause of this association. This is because no actual observation of predation was observed by De Ruiter during the five-month study period. It is difficult to assess whether the lack of predation is a true reflection of the condition, or simply an artifact of the short study period and testing procedure. Altmann (1981) notes the difficulty of determining what and where a baboon is scanning under field conditions, and Rowell and Olson (1983) remarked on the difficulty of determining the precise direction, and more importantly the function, of gaze in patas monkeys even in laboratory setting. Forest-dwelling monkeys are well known for their elusiveness and difficulty of observation (Carpenter, 1964; Struhsaker, 1975), and the technical difficulties in assessing the functions and frequency of "scanning" of these monkeys under field conditions may be insurmountable.

The technical problems aside, De Ruiter's study also raises the important issue as to how group sizes are to be defined. Robinson (1981) reports that cebus species' groups range from 6 to 40 individuals. De Ruiter's "large group" therefore falls in the middle range for cebus monkeys. Clearly, judgement as to what constitutes large or small can be defined only with reference to predation pressure. As predation pressure cannot be measured, it is meaningless to characterize the group in terms of size difference.

The foregoing discussion raises an important question concerning the nature and mechanism of dispersion among neighbouring groups. It has been suggested that mixed species associations among cercopithecus species represent an anti-predator strategy in that the coalition of smaller groups into a larger aggregate leads to increased vigilance and mutual benefit (Gartlan and Struhsaker, 1972; Gautier-Hion et al., 1983). With reference to the two groups in De Ruiter's study, which occupy extensive overlapping home ranges, it would be interesting to ask whether the two groups coalesce in the face of danger. If the prediction regarding vigilance and group size is applicable, would we expect an increased advantage if the two groups coalesce? If such an association is not to be found, what then are the mechanisms whereby group identity and cohesiveness are maintained? These questions relate to the critical issue regarding the nature of grouping and sub-grouping in primate societies.

Primate societies in general are complex and involve different levels of social functioning (Jolly, 1972). It has long been recognized that social behaviour must be studied at several different but integrating levels, including those of individual behaviour, interactions and relations between group members and social structure (Hinde and Stevenson-Hinde 1976: Hinde, 1983). Rowell (1972) points out that some of the data used to support the adaptive nature of social organization (De Ruiter's study being a prime example, see also van Schaik, 1983) is not strictly about social organization at all, but about the spatial distribution and demographic structures of populations. These factors are clearly important components of social organization, but they are by no means the whole picture. There is not a simple correlation between group size and composition and types of social organization. The following comparison will illustrate how a given group composition can result in different societies.

Both hamadryas baboons (*Papio hamadryas*) and patas monkeys (*Erythrocebus patas*) are well-known savanna-adapted species. Both species are described as one-male, multi-female groups (Kummer, 1971; Eisenberg et al., 1971). In the hamadryas groups the adult male is the focus, and females interact relatively little with each other except when one of them has a new infant. The male actively "herds" his females, responding with threats to their straying. Females in turn respond to his threats by approaching him. From the long-term study of Kummer and his students (1974), the hamadryas could properly be described as living in male-controlled "harems." This form of social relation is unique among non-human primates. In the patas, females interact mainly amongst themselves, and the male is rather peripheral in both the spatial and the social sense. Patas males have been described as the most socially detached of primate males found living in heterosexual groups, although they often perform the role of the look-out and protector of the group by diversionary displays. Threats or aggression between males and females are extremely rare in wild patas. Females also direct group movement and regulate intragroup life. They have been observed to coalesce against the physically larger male of the group should he threaten one of the members (Rowell and Hartwell, 1978).

The one-male, multi-female unit is the most stable and the smallest social unit in hamadryas society. This unit has been referred to as a "clan" (Kummer, 1971). Several clans may travel and forage together in the daytime and form a band. And several bands in turn can form a troop, the largest social unit in hamadryas society. Such troops are formed of animals who roost together at night, sleeping at cliffs. Such troops may contain up to three hundred members (Kummer, 1971).

This brief description illustrates that hamadryas society is multi-levelled and interactive. While it has been argued that the congregation of several bands at the sleeping cliff represent anti-predator strategy because of the advantage of large troop size and the limit of safe refuge from predators (Kummer, 1971), it has also been suggested that the pattern of relationship between male and female and the formation of "harems" could better be explained by subsequent competition among animals to increase their reproductive success (e.g., Clutton-Brock and Harvey, 1977). Such a segregation of social functioning is not obvious in patas society. The same group appears to function as a foraging, breeding, as well as antipredator unit (Rowell and Hartwell, 1978).

It is clear from these brief examples that care must be taken in discriminating those factors that favour spatial aggregation or large grouping from those which promote the formation of true social structure (Allee, 1932;

Eisenberg, 1966). It is interesting to note that in a review of predator-prey interaction, feeding strategies and optimal group size, Bertram (1978: 96) concludes that, "in truly social species, the size of the group is probably determined less by ecological factors and more by within-group interactions". Therefore, to describe the social organization of various groups and species so that they are comparable, information on the *patterns of interaction* between members of the same group and members of another group is needed. Such information forms the essential basis of the hypothesis concerning the mechanisms whereby particular social systems arise. To this end, many researchers have turned their attention to the possible importance of reproductive strategies in shaping primate social systems (e.g., Goss-Custard et al., 1972; Crook et al., 1976; Wrangham, 1980).

Ecological Correlates of Mating Systems

A great deal of emphasis has been placed on mating and reproductive behaviour because it is generally assumed that this aspect of social relationships is most subjected to ecological influence, particularly food abundance and distribution (Bradbury and Vehrencamp, 1977; Emlen and Oring, 1977; Orians, 1969; Wittenberger, 1980). The suggestion that resource availability can directly constrain or alternatively promote opportunities for the expression of social relationships, particularly that between male and female, is the subject for the remainder of this paper.

In most sexually reproducing species, particularly in mammals, females invest more time and energy in the production and rearing of offspring than do males. Because of this fundamental difference in parental investment between males and females, it is widely accepted that natural selection will favour an optimal distribution of females with respect to critical resources. The consequence being that female reproductive success is limited primarily by energetic and nutritional constraints. The reproductive success of males, in contrast, is dependent on the distribution of females as well as on the males' ability to defend females against competing males (Emlen and Oring, 1977).

An interesting conclusion from this line of reasoning is that the factors promoting gregariousness for males may be different from those for females, leading to further differences in ranging and dispersal patterns between the sexes. These suggestions have been proposed for a wide range of mammalian and avian species (Emlen and Oring, 1977; Lack, 1978; Orians, 1969; Oring and Lank, 1982; Verner and Willson, 1966; Wittenberger, 1980).

Wrangham (1979, 1980) has incorporated this line of reasoning in his model of primate social grouping. He points out that most primate species follow the typical mammalian pattern of dispersal and residence. Males usually leave their natal group as they approach maturity, whereas females tend to remain in their natal groups through life. Consequently, males may spend most of their adult life with few if any kin after dispersal. Females, on the other hand, tend to be accompanied by close kin and, Wrangham argues, develop closely knit cooperative social networks, which he calls female-bonded groups (FB groups), under certain ecological conditions.

In particular Wrangham suggests that the key factor promoting large, stable groups of females is the advantage they receive from collectively defending resources distributed in large, scattered patches. Under such conditions, females in a group should obtain more food of "high-quality" than do females foraging alone. In contrast, if resources are distributed in small patches that will support only a few females, then smaller, less stable, but more territorial groups are expected. Under such circumstances, females would exclude more than one male from the territory, giving rise to a single-male group rather than the multi-male group, which is expected to be found in association with larger social groups. When group size is large, it may not be energetically feasible for females to exclude the males from the feeding sites. In such cases, groups may become multi-male and intra-group competition between males for females should be expected.

Accordingly, group cohesion is based primarily on female alliance, and males enter the picture as "secondary players" (Wrangham, 1984: 261). Males of course, add to the complexity of social structure, but they do not alter in any substantial way the basic pattern of female relationships. Wrangham's model is therefore in contrast to the traditional view of social evolution, such as that of Eisenberg et al. (1972), which views female characteristics as a byproduct of selection upon males.

Simply stated, then, the factors affecting the reproductive success of the sexes are different. While a male's fitness is limited by the number of females he can inseminate, the fitness of a female is limited by the number of offspring she successfully raises, which in turn is dependent on her success in feeding competition. Male sexual strategy and female feeding strategy therefore interact to produce the patterning and distribution of male and female primates.

In addition to the prediction regarding the formation of single-male groups, two related predictions regarding inter-group encounters can be derived from Wrangham's model: Females in female-bonded groups should actively participate in intergroup encounters. Furthermore, such encounters are expected to be concerned with food, and their interactions should be antagonistic. On the other hand, encounters between males should be related to defence of the female, so that males should resist the immigration of other males and prevent them from gaining access to females in their group. Wrangham (1980) examined data from thirty primate species and concluded that these predictions are largely borne out by the data. Cheney (1983) examined intergroup relationships in macaques, baboons and vervet monkeys and suggested that the patterns of encounters follow closely the predictions of Wrangham's model. The implication of these findings is that relationships within groups cannot be fully understood without reference to intergroup interaction. The mechanisms that maintain stable relationships within groups are inextricably linked to those between members of different groups. Furthermore, it became apparent that what is generally referred to as "territorial encounters" between two groups may be of a different nature for the adult females than for the adult males of these societies.

These realizations are no doubt important, but Wrangham's predictions may in fact underestimate the complexity of intergroup relationships. Interaction between neighbouring groups may be "symmetrical" or "asymmetrical" (Richard, 1985). The former is characterized by mutual antagonism, avoidance, or tolerance.

In species such as Japanese macaques (*Macaca fuscata*), all three of these interactions have been observed between the same two groups at various times (Fedigan, 1982). In an asymmetrical relationship, members of one group consistently replace members of another group. Such displacement may be spatially or food oriented. The direction of displacement, however, is not necessarily determined by the relative size of the interacting groups. Some researchers have therefore suggested that asymmetrical interactions may involve groups that are hierarchically ranked (Dittus, 1977, 1985).

The determinants of particular interactions and the quality of intergroup encounters are not clearly understood. It is difficult, however, to explain much of these interactions in terms of competition for food and sexual partners. The reasons for the difficulties are manifold.

[1] While it is well established that males of many species leave their natal group, the reasons why males "peripheralize" themselves from their natal group and join another are not clear. Wrangham (1983: 261) invoked two popular assumptions:

> An important component of male strategies is that males leave their natal groups to breed, both because of the costs of inbreeding in their natal groups and because of the increased mating opportunities elsewhere.

This statement is accompanied by the suggestion that dominance ranking and mating opportunities, and hence reproductive success, are positively correlated. The implication is that low-ranking males are likely to leave and seek better mating opportunities elsewhere (Drickamer and Vessey, 1973; Parker, 1979).

Long-term studies of Japanese macaques reveal that there is no clear correlation between ranking of males and reproductive success (Fedigan, 1983). The problem of dominance rank and reproductive success will be dealt with in further detail in the "Discussion." Low-ranking males are unlikely to improve their chance of mating elsewhere since they would likely begin with low rank in a new group. Furthermore, there is no clear evidence that low-ranking males are more likely to leave than are higher ranking males. The claim that males leave to improve on the chances of mating is not strongly supported (Andelman, 1986).

[2] There is also no convincing evidence for the claim that male migration reduces the cost of inbreeding in the natal group. Population geneticists have shown that very small rates of immigration by outbred

individuals from different groups are sufficient to reduce inbreeding coefficients (Cavalli-Sforza and Bodmer, 1971; Lewontin, 1974). Reduction of the cost of inbreeding does not call for the high rate of male transfer between groups.

On the other hand, males of neighbouring groups may be closely related in the first place. This is most likely to occur when group fission occurs. When new groups are formed by the fissioning of a larger group, animals of the respective groups may be related or know one another from the time before splitting took place. Fissioning has been observed among rhesus and barbary macaques, common baboons, and Japanese macaques. In all these cases, it has been documented that groups fission along female genealogical lines so that the "daughter" group always consists of matrilineal kin and their allied males (Chepko-Sade and Sade, 1979; Missakian, 1973; Koyama, 1970).

The precise cause and nature of male dispersal and group fissioning however, is poorly understood, partly because of the short-term nature of most primate field studies (Menard et al., 1990). Sugiyama (1976) and Itoigawa (1974) suggest that male monkeys emigrate for various reasons, such as following the example of an older brother or a playmate, conflict with matrilineal kin, or weak bonds with the mother. Males also leave to seek richer food sources and available mates. Indeed Meikle and Vessey (1981) found that males were more likely to move into social groups in which an older brother already resided. However, studies that have been successful in documenting male migration and residence pattern are rare. In the wild, it is often difficult to track the natal males once they have migrated from the natal group. Furthermore, since genealogical relationships are rarely known, it is difficult to assess behavioural interaction in terms of genetic relatedness (Richard and Schulman, 1982; Rowell, 1988).

Dispersal patterns between groups is obviously highly complex and may involve changes in the rank between males and between females, as well as changes in the pattern of alliance between and within the sexes both in and between groups (Furuya, 1969; Koyama, 1970; Maruhashi, 1982; Yamagiwa, 1985). The reason why a male leaves his natal group and his subsequent acceptance into a neighbouring group may, to a large extent, depend on the past history of the two groups. If the groups involved are "sister" groups, then a male, who appears to an outside observer as a "stranger," may in fact be related to the males in the group into which he attempts to migrate. The reactions of the "resident" males to the "immigrant male" may vary accordingly, and antagonism is not the rule. Chism and Rowell (1986: 37) summarized this problem succinctly when they wrote:

> For most species the origin of group adult males is unknown, and since Old World monkey males typically do not mate until they are at least 5 or 6 years old, it is not sufficient, nor reasonable to assume that a "new male" entering the group during a study of a few years' duration is a non-natal male.

Chism and Rowell's long-term study of the patas monkey also points to the danger of assuming that the grouping pattern of a monkey at any one time is the same for the species at all time. Yet such an assumption is a necessary one underlying the ecological explanations discussed thus far. As Menard et al. (1990: 172) recently noted:

> one might ask whether the so-called stability of primate social systems is not partly a result of observational biases. So far, most studies have been carried out in what appears to be the "climax" habitat of the species concerned. In such situations, it is likely that we will underestimate 'the flexibility and variation found within social systems', as pointed out by Dunbar (1988).

It is obvious that while ecological factors such as those identified by Wrangham may influence inter-and intra-group relations, other social factors are likely to play important roles as well. More importantly, unless we have a good knowledge of the "age" and "development stage" of the groups concerned, it is not justifiable to assume the social group to be a stable product of the individual's interaction with the environment.

Discussion

Two major trends have profoundly affected the course of research on primate social organization. Since the mid 1960's, the works of Crook (1965, 1970), Lack (1971), and Orians (1969) on the adaptive origins of avian mating systems have stimulated primatologists to search for similar patterns of interaction between ecology and social systems in primates. In the meantime, development of sociobiological concepts have fostered a broader approach to understanding the evolutionary causes of social variability.

Many primatologists interested in the behavioural ecology of primates, have enthusiastically adopted certain sociobiological concepts. For example, Rasmussen (1983: 96) has stated, "Sociobiological theory seems likely eventually to provide the necessary framework by which studies of primate behaviour and primate social ecology may be ordered". It is therefore not surprising that much of the current debate in primate socioecology is embedded within a broad spectrum of research in evolutionary biology (Richard and Schulman, 1982; Terborgh and Janson, 1986, Crook 1989). While avian models and sociobiological concepts have yielded interesting hypotheses, two of which have been the subject of the present discussion, there are also obvious problems in the application of these models to primate data. These problems are both conceptual and methodological, and unless properly clarified, may have profound influence on the interpretation of primary data and, more importantly, may dictate the ways in which data is to be collected in the first place.

These last points can best be illustrated by further comparison of the predation and food-competition theory. Though different in their conclusions, the two models share a remarkable degree of similarity in that the primary data base are avian species and the same assumption underlies both models: individuals form social groups in order to maximize their own reproductive success. Social organization in this view is implicitly regarded as an epiphenomenon of individual interactions. Social relations are seen, therefore, as consequence rather than cause of individual behaviour. Both the avian models and this view of the individual and society may be inadequate when applied to primates. Several reasons for this claim will be discussed.

[1] Obvious differences exist between avian and primate mating systems. Of approximately 190 extant primate species, only 22 are known to be monogamous (Kleinman, 1977), although there are probably more to add to the list (Terborgh, 1983). On the other hand, close to 90% of bird species are monogamous (Lack, 1968). And while permanent multi-male, multi-female groups are common in primates, they are relatively rare or unknown in birds. Furthermore, leks are unknown in primates (Terborgh and Janson, 1986). Despite these superficial differences, it is more important to realize

that although much of avian mating systems are explicable in terms of control over resources such as nesting sites, territories, and mates, the same can not be said of primates (Rowell, 1987, 1988). The primary reason is that mating behaviour in primates takes place within complex social contexts such that a mating system only constitutes a single component, albeit a major one, of the larger social system that is made up of relationships between the sexes as well as within sexes, and relationships between and within different age-grades (Hinde 1983, Quiatt 1988).

For avian species, mating systems have often been implicitly regarded as synonymous with social systems (e.g., Orians, 1961; Emlen and Oring, 1977; Vehrencamp and Bradbury, 1984). Consequently, emphasis has been placed on male-female interactions and on behaviours that have direct reproductive consequence.

Wrangham and Rubenstein (1986: 455) point out that:

> a focus on intersexual aspects of mating relationships is appropriate in birds because cooperation among breeding adults of the same sex is rare, so that differences among species social systems can be accounted for to a large extent in terms of male-female relationships.

This emphasis on intersexual relationships as the prime determinants of social organization may be inadequate for understanding the complex relations in primate societies.

With few exceptions, the tenure of primate researches are short compared to the generation time of the primates, which could range from five to ten years for a generation for monkey species, and ten to fifteen years for apes. Data on reproductive and mating patterns collected on short-term or cross-sectional studies are not really comparable to data gathered on species that are short-lived and have high reproductive rates. Calculation of fitness indexes based on short term observation can not be extrapolated to explain the long-term consequence of a particular mating strategy (Dunbar, 1979, 1983). This is because the number of offspring produced in any given year will tell us little about differential reproductive fitness of individuals in a given primate population. A particular mating strategy can only be evaluated in terms of reproductive success calculated over the entire reproductive period of the individuals (Altmann and Altmann, 1979; Sade, 1980). At present, long-term demographic data are only known for a small number of species (see Richards, 1985), and the applicability of avian mating models to primates remains to be tested.

The long-term study of gelada baboons (*Theropithecus gelada*) by Dunbar (1979, 1983, 1986) reveals that males employ different strategies to obtain mates, depending on particular demographic factors as well as their development stages. As demographic parameters vary from year to year, mating strategies for any particular male may change from season to season. Dunbar's findings are consistent with the view, long held by Rowell (1967, 1972), that primate populations rarely achieve equilibrium, but are in constant flux due to various demographic processes (i.e., births, deaths, immigration and emigration) (see also Cohen, 1969; Dunbar, 1988; Jorde and Spuhler, 1974; Altmann and Altmann, 1979). These authors argue that such processes may occur randomly and hence are independent of local ecological conditions.

For example, in a small population, a highly imbalanced sex ratio at birth, which can occur by chance, may persist for several generations despite a more balanced sex-ratio at birth in subsequent years. The consequence is an imbalanced adult sex-ratio, which may in turn affect the availability and choice of mates. As most primates live in small populations, the possibility of small sample fluctuations or demographic "drift" are likely (Altmann and Altmann, 1979). Consequently, we may expect population structure and size to change over time. To the extent that social organization is influenced by demographic factors (Dunbar, 1988), we might expect variation in social patterns between groups as well as varying patterns for the same group at different times. It is therefore not surprising that given the short study period, different investigators would describe very different mating patterns and social organization in the same populations. Chism and Rowell (1986), for example, recently challenged the long-held view that patas monkeys (*Erythrocebus patas*) live in an exclusive one-male polygynous mating system. Their long-term study reveals, instead, a fluid pattern of male residence, especially during the times when females are fertile. As long-term studies of various species become available, our picture of primate mating systems may alter substantially. At present, it is clear that the dichotomization of primate mating systems into either "monogamy" or "polygamy," and single or multi-male, is inadequate for characterizing the rich diversity of primate social organizations (Rowell, 1988).

[2] The assessment of a mating strategy is further hampered by the many difficulties involved in the measurement of fitness in primates. The first problem is that researchers often differ substantially in their measures of reproductive success. This problem is more acute with measurement of males. Since direct measurement of reproductive males is difficult, it is frequently assumed that "male reproductive success varies as a function of the number of copulations" (Trivers, 1972: 138). Thus, Bertram (1976), for example, has used copulation frequencies in lions to assess reproductive fitness. Similar approaches have been used to measure reproductive success in primates (e.g., *P. anubis*, Devore, 1965; *M. fuscata*, Hanby et al., 1971; Eaton, 1974; *P. ursinus*, Saayman, 1971; *P. cynocephalus*, Hausfater, 1975). However, the relationship between frequencies of copulation and the probabilities of fertilizing females has not been empirically determined (Andelman, 1986). Furthermore, because of the difficulty of observing copulation and determining insemination in free-ranging primates, many researchers have relied on more indirect methods of assessing reproductive success, such as the number of sexual consorts formed during the conception period (e.g., Carpenter, 1942; Drickamer, 1974; Seyfarth, 1978; Smuts, 1985). The interpretation of what constitutes sexual consorts and social bonds in primate societies is by no means unambiguous, thus adding to the difficulties of interpreting sexual strategy and mating patterns. In short, because of the different methods of data collection and interpretation, measurements of reproductive success for individuals in different populations or species are often not strictly comparable.

There is also the added problem of determining paternity. For primates, as well as for other mammals, the common assumption is that the harem male in a uni-male polygamous group is responsible for most conceptions within his group. In multi-male groups, it is generally assumed that there is a lineal relationship between the probability of insemination and male dominance rank. Evidence for these assumptions are highly equivocal.

In a comprehensive review of dominance rank and reproductive success in primates, Fedigan (1983: 124) arrived at the following conclusion:

> The answer to the deceptively simple question, 'does high dominance rank in primates lead to greater reproductive success?' has been variously given as "yes," "no," or "in part." To date, the general answer should be that we really do not know.

The aforementioned problems bring into question the utility of characterizing, let alone classifying, primate societies in terms of mating system.

[3] Preoccupation with ecological correlates of mating strategies have left us with little understanding of the variation in the relationship between social and ecological variability among individuals of different age and sex classes. As individuals differ in age, physiological state, and social status, their ecological requirements may differ substantially. Mating strategy is a critical component to such variation, but such strategy alone does not account for the complex relations between age-grades and within the sexes. What, for example, is the relationship between age and the quality and quantity of diet? How do intrasex relations affect access to preferred food sources? Little is known about these relations in primate societies (Richards, 1985; for an exception see Post et al., 1980), perhaps due to an overemphasis on feeding behaviour only as it pertains to reproductive success.

The disproportionate concern with mating strategy may also undermine the complexity of social roles in primate societies. While attention has focused on reproductively active adults, sexually immature individuals as well as those past reproductive age also play important roles in the maintenance and survival of the social groups. Primates are highly flexible in their behaviour. This high degree of plasticity can be largely attributed to the long maturation period during which behaviours are learned and continually modified. For example, the recognition and choice of mates, as with dietary choices, is not simply a direct result of past selection, but to a large extent depends on the individual's past experience as well as the social tradition that is already in place (Burton and Bick, 1972; Galef, 1976).

In her survey of twenty-one cercopithecoid societies, Burton (1977) notes that the variance in sex roles within and between species is far greater than that predicted by models of mating types. All macaque species (such as the rhesus, *M. mulatta*, the bonnet *M. radiata*, the barbary ape, *M. sylvanus*, and the Japanese macaque, *M. fuscata*), are polygamous, but the range of relationships within and between sexes is enormous and poorly correlated with mating patterns. Similarly, both baboons and Japanese macaques are polygamous. But while female baboons tend to consort with males with whom they have previous friendship bonds (Hausfater, 1975; Packer, 1979; Smuts, 1985; Strum, 1987), female Japanese macaques tend to exclude males with whom they have year-round friendship bonds as mating partners (Enomoto, 1974, 1978; Fedigan and Gouzoules, 1978). The relationships between males within primate societies are equally variable. As Chism and Rowell (1986: 38) recently point out:

> Male competition does not take the form of rare brief contests for residence in a group with the winner automatically gaining a large increment in reproductive success. Male interactions are not limited to competition for access to females. Males live with other males for long periods, in relationships which are presumably mutually advantageous, yet at present we know little about the dynamics or functional significance of male association.

It has generally been assumed that the amount of parental investment is strongly correlated with mating types so that parental investment by males is minimal in a polygamous system and highest among monogamous species (Hamilton, 1963, 1964; Maynard-Smith, 1977). Again there are many exceptions to this generalization in primates. For example, while adult males of bonnet (*M. radiata*), pigtail (*M. nemestrina*), and rhesus (*M. mulatta*) macaques are generally indifferent to infants (Breuggeman, 1973; Simonds, 1974) male barbary macaques play a substantial role in the care and socialization of infants (Burton, 1972; Deag and Crook, 1971; Taub, 1980). Male care of infants has also been reported in Japanese macaques, but the frequency and the intensity of male care-giving vary from group to group (Itani, 1959; Alexander, 1970; Fedigan and Gouzoules, 1978). Such difference in male infant-care behaviour has been attributed to differences in the social tradition between groups (Burton and Bick, 1972; Fedigan, 1982; Burton, this volume). As parental care is the primary process of socialization, it may exert important influence on an individual's subsequent mate choice. Therefore differences in local history and social behavioural tradition must be taken into account in order to understand the diverse nature of primate societies.

The importance of local social tradition in shaping the patterns of relations within societies therefore undermines the deterministic role of ecology in shaping social organization. Several decades ago Allee (1931, 1938) emphasized the importance of the social group as the repository of social and biological information. In long-lived social primates, group identity often endures over time and over generations despite regular change in personnel (Mason, 1976, 1978; Rowell, 1972). This important characteristic of group life is conspicuously lacking from recent theoretical discussions of sociality. The mechanisms whereby group cohesiveness and permanence are maintained still remain poorly understood (Hinde, 1983; Cheney et al., 1986). A better understanding of these mechanisms and the ways social tradition influence the behaviour of group members should form the necessary background for the further investigation of the interaction between ecology and social organization.

[4] The current one-dimensional view of the individual as the locus of selection and its attendant emphasis on maximization of inclusive fitness adds little to our understanding of the interaction between social behaviour and ecology. Indeed this view of animals as selfish competitors and gene machines creates a totally distorted picture of animal social life. A large group size may reduce the chance of one being preyed upon; it is difficult, however, to conceive the loss of a group member to a predator as a possible benefit. If the prey victim is a group member with whom one has close social ties, then the social cost may in fact outweigh the biological benefit of not being eaten. The loss of a group member in primate societies often has social impact on the group as a whole. The fact that individuals form varying degrees of social bonds with other members of the group confounds the measurement of cost and benefit ratio in simple genetic terms.

As pointed out earlier, the treatment of resource acquisition exclusively in terms of competition is equally problematic. Given that resources are distributed over space and time, and that individuals are well aware of their positions within the group, there is always room either to avoid competition or engage in tolerant interactions (de Waal and Yoshihara, 1983; Burton, 1984; de Waal, 1986, 1989). The need to tolerate other members of the group was seen as central to the evolution of social systems by Eisenberg et al. (1972). This view has since received little attention, but other aspects of their paper, such as the advantages of a large group, and intersexual competition, continue to be well cited. The study of social organization is not complete without explicit attention to the ways animals make use of various options to ensure the proper functioning of the social unit. Behavioural scientists have long been interested in information processing, learning, problem solving, and even social manipulation in animals. But until recently, there has been relatively little attention paid to the mental or cognitive experience that underlie much of social behaviour (Griffin, 1976: 1981). Primates in general are unique amongst mammals for their cognitive capacity, and inferences concerning their ability to willfully manipulate social relations to achieve biological and ecological demands further illustrate their cognitive awareness of the social as well as the physical world in which they live (Kummer, 1981; Mason, 1981; Cheney and Seyfarth, 1982; Burton and Chan, 1987; Cheney et al., 1986). How primates make use of their social and ecological skills is an area that is still poorly understood and further research on this topic is clearly needed (de Waal and Luttrell, 1989).

[5] Social organization is by nature a group phenomenon and is not readily reducible to the level of the individual. The lineal causal relationship between ecology and social organization postulated by models discussed in this paper are inadequate to account for complex systems. To comprehend social systems, a multi-dimensional theory built on hierarchical conceptualization is needed. Sewall-Wright (1931) championed such an approach to the study of the multi-levelled nature of biological organization. Recently, an increasing number of evolutionary biologists (e.g., Gould, 1982; Gould and Lewontin, 1979; Vrba and Eldredge, 1984) have called for an expansion of evolutionary theory that would encompass multiple processes, in addition to natural selection, operating at several hierarchically integrated levels.

An important characteristic of this conception is that a simple, lineal, causal relationship between lower and higher levels is no longer tenable. Instead, higher organization can exert what Campbell (1974) has called "downward causation" in which "all at the lower levels of a hierarchy are restrained by and act in conformity to the laws of the higher levels" (Campbell, 1974: 180, cited in Hendrich, 1983: 748). A possible implication is that social organization may constrain and define the individual's environment such that individual behaviour is not necessarily a direct response to ecological conditions.

For example, an individual that is made peripheral by other members of a primate group is more vulnerable to predators. Could the fear of being excluded from the group be a means of social control exercised by the group? In this case predator pressure can not be measured without reference to the process of social interaction. Such a view necessitates alternative explanations of ecological factors and social behaviour, such as the correlation between time spent on scanning the environment and group size. Could members of a larger group spend less time scanning because more time is spent on interacting with group members? Similarly, could members of a smaller group scan more frequently and longer because they are searching for social contact, since the number of social partners is limited in a small group? The same argument can be applied to food acquisition such that if food choices can be seen as socially defined and regulated, what constitutes scarce or preferred resources can only be understood in terms of the interacting social system.

Another important implication of the multi-levelled interacting view of complex systems is that it allows for more than a single steady outcome. It is currently popular to view social systems as products of natural selection which are then dubbed optimum "solutions" to "problems" posed by the environment (see Emlen's quote above). The fundamental problem with this view is that it is neither possible to deduce the "solution" from the "problem" nor is it possible to infer the "problem" from the "solution." To do the former is to omit the fact that there are multiple solutions to the same problem (see Clutton-Brock and Harvey, 1978 and Rowell and Olson, 1983 for primate examples). To do the latter, however, is to commit the most fundamental error in logic of affirming the antecedents: similar consequences do not necessarily imply a similar cause. In short, the use of simple lineal logic is problematic.

The hierarchical view is an attractive alternative, since a multidimensional system with non-lineal relationships guarantees that multiple social organizations, or "adaptive peaks," to use Wright's term, are possible for any given set of ecological conditions.

Conclusions

This paper has examined what at first glance appears to be a well-defined and rather straight forward issue - the explanation of primate social organizations in terms of specific ecological variables. It was argued that dealing with ecological constructs requires treating a broader issue - developing a means of explanation that takes into account the multilevelled, and dynamic nature of social behaviour as well as the non-lineal relationship between the environment and social organization. The inherently complex nature of primate behaviour guarantees that no single theory or set of concepts will likely be applicable to all species.

More importantly, theoretical constructs are useful guidelines within which research may be organized. They are, however, no substitute for observation. Many assumptions underlying models of grouping patterns and mating types remain to be verified, and the testing of these assumptions may well provide the direction for further research. For the time being, it is clear that more long-term studies of a wider variety of primate species are needed for the assessment of ecological explanations of social evolution. As Hubert Hendrich (1983: 749) has remarked in a similar context: "too much over-simplified theory keeps the student (who is in danger of losing contact with a natural environment anyway) from observing animals and nature long enough to begin to recognize the structures and their relations." Unless we come to know intimately the world in which animals live, it is simply impossible to identify the "problems" that are relevant for the understanding of their behaviour. As primates differ in their cognitive capacities because of their diverse phylogenetic, historical, and social heritages, the perception of what constitutes "problems-out-there" will necessarily differ. The idea that animals, regardless of their phylogenetic background, resolve "similar ecological 'problems'" in terms of "a predictable convergence in their 'solutions' as shown in their social organizations" (Emlen, 1980: 129) is simplification of the richness and diversity of animal social behaviour. Non-human primates have been most extensively investigated in terms of their cognitive abilities and social relations (Rowell, 1972; Kummer, 1981; Mason, 1981; Burton, 1984; and the various papers in Byrne and Whiten, 1988). Much of this knowledge, however, remains to be integrated into studies of primate ecology and behavioural ecology in general (Jolly, 1985; Cheney et al., 1986; Terborgh and Jason, 1986; Rubenstein and Wrangham, 1986). It is evident that while there has been a steady importation of concepts from sociobiology and behavioural ecology, the opposite flow of ideas is much needed. The term "socioecology" is misleading unless the social components are fully appreciated in ecological studies.

CHAPTER TWO

The Social Group as Information Unit: Cognitive Behaviour, Cultural Processes

by

Frances D. Burton

Introduction

This chapter is about the social group as a unit of information. That information is of two kinds: it is biological, transmitted through reproduction by DNA (genes) and it is social, transmitted through reproducing of behaviour patterns (i.e., learning). Non-human primate communication can be viewed as the individual projecting images created by its self. These images, or representations code meaning, some of which is exclusive to the local group. Because non-human primates are cognizant of the reactions to their actions, they can manipulate their representations for a given outcome. Manipulation of representations, however, depends on commonly understood or more precisely, shared conventionalized meanings. Since these conventionalized representations extend over time within the local group, it is possible to recognize local group tradition, as well as the process by which tradition changes so that information is constantly renewed.

The focus of this chapter is on monkeys, primarily macaques, which are the monkeys I know best. The "facts" of the matter are occasionally juxtaposed to narratives. These are intended to allow the reader to depict the scene, to *actualize* the facts, to recreate my experience which gave rise to theory.

> Ben ambles into the open area at Caroline's Battery, crosses to the periphery and hops onto a stone bench. He rears onto his hind feet, head-up, scanning the horizon. He is intently listening. His pinnae move on his head, gathering the sound into his ears, his eyes are bright, his overhanging brow ridge creates a scowl. His concentration is contagious, so that some young ones nearby pip with repressed fear calls, bodies tense, mouths drawn back. They run to the shelter of their mothers. Some older females look at Ben. They sense his tenseness, and look around. They gather their infants onto their breasts and shuffle away. A fight breaks out in response to Ben's tension; the *wraahing* sounds intensify the mood. The hackles at Ben's neck rise; his facial muscles tighten still more. He gives a low throated bark; the young ones jump, climb, hop into the trees; some females move out of sight; some stay near him, looking in the same direction. They too bob up and down, rearing onto their hind legs as Ben has done. Two adolescent males run ahead, in the direction of Ben's gaze. Now I see it: a dog on his lead approaches. The monkeys watch closely as pipe-smoking master and frolicking dog make their way past the Battery, along the road. Ben gets off the bench, flops down before a female friend, and she begins to groom. At this signal, the group of barbary macaques goes back to less tense activities: feeding, grooming, sleeping while the young play.

Witnessing such interactions over and over again, reveals the profound integration of the non-human primate social group. Information one animal receives is passed throughout the group by looking at, by vocalizations, by hackles raised, by body motions. How one reacts depends on who one is, how many relatives one has, and who they are.

Monkey Society Defined

A society is more than an aggregate of individuals. It is [1] characterized by being bounded in space - members of the group are integrated; relating to and related to each other more than to individuals beyond that demarcation. A society is [2] boundless in time, as the social group continues beyond the life of any member, so that it is characterized by [3] having generations: minimally, breeders and young. Primate societies exceed that minimum, with a post-breeding generation. A society is characterized by its members exhibiting [4] "a degree of cooperative action based on their ability to communicate with one another" (Alcock, 1975: 403). Finally, a society [5] is an information unit, comprised of all the activities that have been stored by its membership. Storage is genetic, in the individual's DNA and social, in the individual's memory.

A non-human primate society is composed of individuals of different ages, and, generally speaking, for most of the year, of both sexes (there are some exceptions). Tenure in the group is variable, depending not only on the longevity of a species, but also on social process. In many groups of non-human primates, for example, one or the other sex leaves the natal group as sub-adults. Generally, however, the group will contain members of different generations: infants, newly arrived - the product of complex biological activity as well as patterns of friendship, courtship and mating - juveniles, sub-adults, and adults at various reproductive and social stages.

Each generation is constrained by social norms. Roles and frequencies of activities vary from age-grade to age-grade. Infants, for example, will not be protectors; adults are generally not playful. Only adults are reproducers, and of adults, most of the reproducing is done by prime and mature adults; old adults may live past reproduction. Yet **whatever** the reproductive function, all the age-grades have social roles.

Infants are 'social adhesive.' The infant is born into a matrix that literally contains and centres on its mother, but is peopled with siblings, as well as lateral and lineal kin. The infant is attractive to the group, is sometimes passed around, is usually the object of soft stares, grooming attempts and general proximity behaviour. There is much vying for a chance to get near to the infant.

> Charlotte sits with her newborn. She is weary, still streaked with blood, the umbilical cord dangles from her infant. She is approached from behind, the approacher, a female, makes cooing, crooning low-pitched noises so that, perhaps, her approach and person is known. Charlotte turns her head as if in acknowledgement, and tiredly lip-smacks. The female lip-smacks too, and begins to groom. Slowly, her hands wander across Charlotte's back, slowly, towards the infant. Charlotte knows this female; has lived with her for over a decade. She is not stirred. The female continues her slow, almost stealthy groom towards the infant. Reaching it, she grooms a caress,

> but Charlotte breaks the contact with a slight move of her shoulder. The female quickly resumes grooming Charlotte's back.

The juvenile is at the first stage of independence and so at the beginning of learning social responsibility. Most of its time is spent in play, but play is not aimless; it is the work of the young. Play, being (in a general sense) exploration in a protected environment, allows for free variation. Through play the juvenile learns about itself, about the capabilities of others, about its relationships in the group. In addition, the juvenile, largely through play, develops new ways of doing things, even using its body in new ways like walking on its hands or covering its eyes; or discovering new foods or inventing new ways of preparing them. Play encourages flexibility as well as "certain aspects of learning ability" (Fagen, 1981: 476). The memorization of these patterns ensures an enlarged repertoire of motor movements, behaviour patterns, gestures, and the like which can be called upon under novel circumstances.

Juveniles are learning elementary adult behaviours. They may babysit infants and certainly most exuberantly give warning barks.

> The crows are raucous this morning. Several swoop down from the peak. Three juveniles chorus a warning bark at their flight, but the adults know these birds and do not respond. The juveniles do not stop barking; rather they sort of wind-down.

The responses of the older monkeys inform the juvenile as to which situations are worthy of warning, thus training the juvenile to discriminate appropriate stimuli from unimportant noise.

Sub-adulthood is a bridge between child and adulthood. These adolescent monkeys are caught up in the tension between playing as a young one, and assuming adult behaviour. Capable of reproducing, the sexual movements they have been practising since infanthood have new significance. But they are not allowed to participate. Social strictures by adults curtail their sexual maneuvers.

> Jake is slim, his long limbs add fluency to his swagger. He is up on a rock when I first notice him, sunning, one lithe limb casually draped over the side. His eyebrows twitch, his pinnae move; he is only apparently casual. Every sense is taut. He is watching; waiting. The object of his attention appears over the edge of the wall below him. She is full-grown, a new adult. She hesitates at the top, then springs lightly to the ground. He

> is up. Looking in every direction, he jumps to the roadway. She stands, and presents to him. He rushes towards her, and begins to mount, placing his feet at her ankles, his hands at her hips. Suddenly, an adult female is there. She threatens him, forwarding her shoulders, extending her mouth, her eyebrows raised. He cowers, and runs off.

Adulthood is the longest period in a monkey's life; and monkeys can live to 25 years. It is not, therefore, a homogeneous stage and different field workers recognize different divisions of this period. Minimally, there are young, prime, mature and old adults. Young adults have a great deal in common with sub-adults, but are emerging from that transitional stage. Depending on species, these young animals are adult at from 3-6 years (earlier for *Cercopithecus* and *Macaca,* later for *Papio)* and attain this stage when they are socially permitted to begin their reproductive and parental lives. With time, adults put on bulk, and are responsible for most of the breeding. They become increasingly more conspicuous politically (Mason, 1986). Mature adults are socially and politically prominent, even though as maturity advances, reproductive activity diminishes, while pigment spots, greying, growths, and extra bulk appear. Descent into 'old' is noted by increased fragility, arthritis, tooth loss, deterioration of vision, and marked decrease of reproductive activity. Old monkeys, however, retain a surprising degree of social and political vitality.

> Bridget can barely see. Her brow ridge is so overgrown, it flops in front of her eyes. Like a person with bifocals, she must tilt her head back to get a proper view. She has trouble with sunflower seeds because her incisors are worn down to a nubbin, but she enjoys sitting in the sunshine, cracking them in her molars, working the shells out of her mouth with her tongue. She is approaching the slope where the sun is now full and strong. Under the single tree, whose shade is most attractive, lie a young adult female, her infant, a juvenile, and nearby, a sub-adult male. Bridget sits and glances. Her mouth sets into a firm position and she glowers at the resting group. Getting up, she shuffles in her arthritic but resolute walk towards them. The juvenile sees her first. He starts, and then jumps into the tree, alerting the others who sit up abruptly, take in the advancing, determined Bridget and leave. Bridget has reclaimed 'her' tree.

Note that the active reproductive function of the animal occupies only a portion of a life: from (on the average) 4 years until about 15 in a lifespan that may stretch to 25. Is all that extra time (energy) about ensuring which (genetic) template succeeds? Or is monkey life about sociality of which reproduction is merely a facet? If the purpose (plan, design) is about reproduction, what is life once the purpose is fulfilled? Clearly there is more content. It is sociality which gives the definition (significance, meaning) of life for gregarious primates. The actions, reactions and interactions of the members of the social group, categorized as friendships (Smuts, 1985), political maneuverings (de Waal, 1982; Mason, 1986), sexuality, nurturance, education - whatever - establish the content of non-human primate life. These constitute a network, a *plasma* which both binds the membership into an integrated unit and informs the members as to their relationships and roles within that unit.

Information Transmission

Non-human primates differ from other mammals in several ways. They share with cetaceans and elephants a fetal brain growth trajectory that differs from all other mammals (Deacon, 1990). Relevant to the discussion here is the fact that the increased lifespan of primates is considered to be directly related to the amount of brain tissue. The ratio of brain weight to body weight at birth is typically 6% in mammals. In primates, however, it doubles to 12%.

> This growth pattern probably evolved at the beginning of the strepsirrhine radiation in the mid to late Paleocene, initially as a means to reduce maternal reproductive investment, in adaptation to a requirement for precocial birth in a strictly arboreal niche, coupled with a relatively restricted insect food resource. When a wider range of plant food resources became available, and body size increased, the Haplorhini, including Tarsius, *retained* the 12% trajectory while returning to a high level of maternal investment. (Sacher, 1982: 97-112.) [italics mine]

Increased care-giving, and a longer life span coupled with a large brain with an enormous storage capacity, identifies the distinctiveness of primate existence. The products of social interaction, that is, the patterns of behaviour are witnessed and memorised. Genetic products provide raw materials including rapidity of nerve impulse, emotional states, muscles, neurons, developmental schedules. Response to these products defines the self, and interaction between "selves" creates social behaviour. In turn, social behaviour affects emotions, hormones, etc., in the self as well as in others, and so, a dynamic system is described.

Primate societies stand out amongst mammals generally in having time depth. A non-human primate will likely have a living grandparent, and in some protected environments, a great-grandparent as well. All these relatives contribute to the society as information unit. With several generations present at the same time, the amount of information and variety of it becomes virtually infinite.

Information is transmitted from generation to generation in at least two ways: biologically by the transmission of templates for protein synthesis and socially through the assimilation of patterns of traditional behaviours. Conventionally discussed as dual, even opposed functions, in reality these constitute a singular process - development. This is a complex, unified process which engages genetic products in the individual's progression within the context of its ecological and social environment. It is therefore that the organism has its individual, ontogenetic history and its group history by the moment of birth. This explains both the similarities the neonate shares with others of its kind and the differences unique to itself. Development is a dynamic process; it is highly responsive to information received at any moment. Templates can only operate on the information or resources they receive. Cascading events are governed by the paths or channels they occupy - and - whatever gets in them.

> The rocks in the stream near TaiPo road, thick with trucks and cars in the all-day rush hour, were darkened with gasoline that had been poured to control mosquito larvae. "Little Mama," nicknamed for her activity that breeding season, came down from the branch overhanging the road and drifted towards the stream where she jumped onto a rock, and lowered her mouth to reach the water. When next we saw her, some months later, she cuddled a infant whose hands and feet were malformed into "lobsters' claws."

Transmission of information has both radical and conservative modes. In biological transmission, mitotic replication ensures the conservation of patterns that, assumedly, have been tried and tested. Recombination, on the other hand, which occurs during the process of meiosis, as well as mutation, provides 'new' information in each generation. While it may be interesting to explain a variant of a trait, *variation* itself is a given. Social transmission of information is analogous. Societies have older generations which guarantee the perpetuation, through reinforcement or example, of patterns that work.

> Mark ambles towards Wilma and hunkers down beside her. He places his left hand on little Rosemary at Wilma's teat, and his right hand on Wilma's back, repeatedly clutching her fur. He is "nattering" - opening and closing his mouth with quick tongue flicks and drawing of his lips back and forward - with Wilma at little Rosemary. Each time the infant, pulling away from the teat, makes a sucking movement, which is too pursed, but at least has an open-and-close-the-mouth component, Mark and Wilma repeat their rapid natter. By the third day of her life, Rosemary returns a natter when it is directed at her. Her expression of the pattern is awry, the muscles not yet controlled, but her intent seems clear: she only clomps her mouth open and close in response to the social natter.

Learning

Monkeys educate by modelling. The purposive or willful instruction of motor behaviours or the inculcation of manners assumes an agent who *actively* teaches these things. Since the motivation to teach cannot be ascertained in non-human primates, it must be assumed that it is not present; that *teaching per se* does not exist in prosimians and monkeys. As with Rosemary learning the social natter, activity of the young can be stimulated, directed and reinforced so that it comes to approximate adult behaviours. Monkeys are active learners, but apparently, not teachers. They imitate, that is they observe, remember, and interpret what others do. The stored knowledge then becomes part of the individual's repertoire, available whenever appropriate. Imitation, means duplicating or copying. It is creating a mental image, a representation. The process of imitation is subjective; the individual absorbs and recreates an image filtering through its 'self.' The 'self' is an amalgam of life history to the moment: development, genes, experience, and whatever. Because all the components in the developmental process of one individual can never be identical to those in another individual, the 'self' must be unique. It follows then, that two witnesses of an action will have slightly different versions of the same thing.

The timing of the appearance of the learned behaviour is not predictable, because absorption of the pattern may not be observed by the researcher. It is therefore that early ethologists proposed ready-made behaviours (Fixed Action Patterns) that would appear in full detail at the suitable stimulus (Trigger Releasing Mechanism). This was a lock-and-key metaphor of stimulus-response. But watching monkeys watching monkeys gives the observer another option. Since non-human primates have vast storage capability, the time span between behaviour observed and re-enactment of the remembered image, can be very great indeed. The witness of the process of imitation suggests this mechanism.

> A sub-adult stands looking down the road. He looks down the road, then up, then places his right hand over his eyes, and jauntily ambles in a curious three-legged stride down the road. A juvenile has been watching from the vantage point of the wall. Once the sub-adult is gone, the juvenile takes his place at the top of the road; covers his eyes, and in that odd gait, follows the leader.

Information and Cognition

The imitation of motor patterns encourages physical development and may extend the repertoire of motor movements. The absorption of abstract patterns, however, has also been attested. Cognitive maps are mental representations of the environment. In Gibraltar, the once-a-year trek to retrieve ripe prickly pears, some distance off the usual forage track was led by an old female sensitive to the environmental cues that suggest the immediacy of fruiting time and proceeding, apparently, according to a mental representation of the destination.

> The young juvenile is bopping alongside its mother, rather distracted on this hike by unaccustomed sights. This is a route I am told the monkeys take only once a year, when the prickly pear is in season. An older female leads the way: Ben, the new leader comes after, with a mixed-aged bunch. The progression takes them a distance from their usual area. They follow each other in a wide swathe, but with no particular logic; everyone seems to know where everyone else is. I wonder what cue they are using to know the pears are ripe and wonder about how many others also hold a variant of this mental map. The juvenile has kept up with his mother, taking in, it seems, road signs to mark the way.

Investigations of hamadryas baboons *(Papio hamadryas)* indicate that foraging is done with a representation of the environment - a map (Sigg,

1986; Sigg and Stolba, 1981). Cognitive mapping has been observed in apes (e.g., Galdikas and Vasey, this volume) and has been experimentally affirmed in higher primates by Menzel and his co-workers, and is sufficiently sophisticated to permit translation of visual information about the location of food objects from a two-dimensional monochrome screen to the three-dimensional living area (Menzel, 1991, 1979, 1978; Menzel and Juno, 1985; Menzel and Wyers, 1981). Representations of environmental information are socially communicated - either indirectly as when a monkey directs troop movement, or directly as when, in Menzel's experiments on this subject, a chimpanzee informed his group as to the exact location of some food object (Menzel, 1971). Communication of environmental images even has semanticity - originally found in vervet monkeys, where Struhsaker identified discrete calls for danger from different sources (Seyfarth and Cheney, 1981; Struhsaker, 1967.)

Yet having emphasised that learning - even of complex information - proceeds by observation, storage and imitation (Hall, 1963; Itani, 1965; Jolly, 1966; and see Galef, 1976 for a partial review), qualification is due. A mother removing food from an infant's mouth or hand, distinguishes for it what is edible; a male tapping a female on her back, pushing her towards her infant as if to get her to pick it up in the presence of another monkey, seems to be cueing her. Young males, sensing the arrival of an intruding animal look to the leader male with a warning 'wraah' as if to inform him: Some information is apparently conveyed with intent. Rosemary again:

> At the tender age of one week, Mark takes the infant Rosemary and places her before him on the ground. He moves backward away from her, about half-a-meter away. Lowering his head to look right at her, he natters to her. Only a few days into her life, little Rosemary is conversant with this positive social statement, and returns the natter while making incompetent crawling motions towards Mark. With a gait somewhere between a pull and a hop, she struggles towards him. When the gap between them is slightly narrowed, Mark steps back and Rosemary starts to crawl towards him again. With any interference from a juvenile, or solicitous individual, Mark scoops the infant up, nattering all the while.

The time depth of three or four generations conserves prior knowledge, and ensures a large behavioural repertoire. The older animals are the repository of information that has been tested. Their remembering of tried-and-true patterns ensures a repertoire suited to the particular situation of that local group. Still, the young cannot learn all they need to learn from others. Even amongst humans who rely on stored information that is transmitted without agent, that is, the written word, much learning comes from individual discovery. In the exercise of their bodies, young monkeys discover their capabilities and limitations. They adopt strange postures: walking on the backs of their hands, or with one limb off the ground; they perform radical behaviours: leaping off nine meter trees into water; they utter novel sounds, or sequences of sounds; they explore their surroundings: investigating insects, poking at holes; and they explore relationships: wrestling with each other, approaching infants to touch or attempt to carry, playing at sexual contact with monkeys older and as often, younger than themselves (Bruner et al., 1976; Fagen, 1981). Information within a society changes, even mutates from such rediscovery. New patterns emerge through the play behaviour, discovery, experimentation and interpretation of the young. Enactment of remembered patterns however, is filtered through the individual's being. It is to be expected that each enactment will be different to any other, as, by analogy, pronunciation of words gives phonetic variation without altering the *meaning* of the word. The variations in pattern are what makes information dynamic. Each transmission transforms the information (Sperber, 1985).

> Donald is not the brightest of vervet juveniles. We know this from watching him with Debbie. She has invented a swing game, employing humans and Donald can't follow. First, she climbs up the human, and pushes one arm, so that the perceptive human will extend the arm. Next she sits, then slides backwards so that her knees grip the arm, and she crosses her arms across her chest, gripping her shoulders with opposite hands. The human is supposed to sway the arm, so that Debbie gets to swing. Donald tries. He gets as far as climbing the human, and is helped into prodding the arm into place. He gets into the right leg position, but he can't deal with both arms crossing so one crosses, the other hangs down. The swing-through is therefore imbalanced and Donald gets irritated, scratches and runs off.

Representation of Self

Observation, imitation and interpretation, is the way non-human primates *represent* themselves to others. Representation means the creation and enactment of images. Image is used here in a broad sense to include feeling states and actions, as when - in the non-human primate context - the self conjures an image by moving in a particular manner towards another individual. Enactment stimulates the recipient viewer to remember, recreate, and interpret these images, to produce, as Rumbaugh and Pate suggest, "...a new topography in the pattern of behaviour" (1984: 570). The storage and recall of innumerable environmental images is the hallmark of non-human primate mental activity and, according to Griffin (1976; 1984) a necessary condition for awareness.

At whatever mental level, non-human primates are aware of their representation of themselves, as well as of the representation of group normative behaviour. Representation of self in non-human primates has two aspects: on the one hand, it is the construction of an image for projection. There is however, a more passive aspect which is the reflection of oneself without manufacture. This is termed self-image and reveals how an individual feels about itself; how it is prepared to engage others, to interact. Self-image in non-human primates is testable. Mirror experiments have shown that the great apes respond to their reflection as a representation of their selves (Gallup, 1970; Suarez and Gallup, 1981; Patterson, 1978). They recognize the form before them as unitary with their sensibilities. Monkeys, however, do not indicate a similar ability under test conditions (Anderson, 1984; Gallup, 1982; Gallup, 1970). Does this mean they do not convey an image of self (Gallup, 1982)? Perhaps not. In experiments with macaques using mirrors, I was taken with the nature of their response to the viewed form. The form was a representation of another monkey of a particular age/sex/status group. Since the field of view was circumscribed by the mirror itself, the reaction of the monkey was specific to the reality s/he perceived. Hence what was being tested was the way the subject reacted to a kind of individual: a young adult male seeing a young adult male, reacted with fear grimace and appeasement gesture; a sub-adult male, responded to a sub-adult male with a play face; one old adult female responded to her mirror-image with lip-smacking; another equally old, with a threat. Clearly what was seen was a representation of an age/sex category: not an individual. The viewer had no visual antecedent relationship with the unknown image. The reaction, therefore, literally reflected the viewer's engagement to 'the stranger,' hence, what was being tested was the representation of self the viewer projected in that second aspect: the reflection of self without construction. Confirmation of field observation comes from laboratory experimentation of this capacity. Yoshikubo prepared slides of rhesus monkeys grouped into classes based on shared characteristics. Results suggested that the monkeys were able to conceptualize a "...class of not rhesus monkey" (1985: 298), that is, to identify their own group as an entity distinct from a group of another species.

But the non-human primate perceives the reaction to its action (Burton, 1984; Cheney and Seyfarth, 1990; Seyfarth, 1987; Shafton, 1976). That information becomes part of a pool of data concerning the image that was projected whose elements include: the animal's self, the receiver's reaction to that self as projected, as well as the elements of the representation that elicited the reaction. However mentally processed, then, the non-human primate has become aware of the projection, and can manipulate this representation to manufacture an image. There is a segment of film, for example, on the baboons of Gombe (Swan Productions, 1975), where a young adult male, 'The Stranger,' having moved in on the local group, is backed into a lake by two young males. He has nowhere to go. Another step backward, and he is clearly vanquished; his bid for place in the group is lost. Instead of moving back, however, he leans forward. His repositioning alters the projection of himself, and the two young males back away. The 'Stranger' can stay. Similarly, even known individuals can become "strangers" by recreating the image of themselves. This is what occurred when an adult female macaque from Gibraltar threatened a subadult male. In the ordinary course of events, he would have lip-smacked in return or fled from her. Instead, he threatened her; he was tense and his threat was tentative. The adult female however, reassessed the image and backed away.

In some fashion, his holding his ground, his projection of a self at variance with her expectations, recreated a being which required a new response.

Sometimes, the recreation of one's representation requires drama. The classic case concerns the chimpanzee, Mike, and the tin gasoline drums. Recasting his image as a noisy, awesome male, produced a new representation and thereby permitted him ascendancy to a new role (Goodall, 1986.)

Acknowledgement of the immediate consequence of an act may well have been the historical basis for the evolutionary capacity to make constructs of futurity; reflexive capacity, once

> ..."gotten from the 'inside' to the 'outside,' that is, transformed from solely mental images to externally visible signs made in behaviour or sounds that broadly represented individual reflexive products, then an evolving primate population would be able to share quickly and widely among themselves the several versions of the consequences of present and past behaviour for the future of not only solitary individuals but the entire population (Williams, 1972: 240-241).

Yet non-human primates do not acknowledge evolutionary time. They exist within personal time; they live within bounded time and space. Time is yesterday whose memories are stored; today is sensible: what is currently known, felt, experienced. Tomorrow is likely not an abstraction for them. At best, the future exists as a goal currently held. Seen after the fact, it appears that a non-human primate has developed a complex strategy over time to effect a purpose: there is doubt that this is what actually occurs. Rather, the memories evoked by an immediate image prompt the animal to perform. Thus, in political machinations, the monkey acts today for an outcome or goal that is prospective; but the moment of its occurrence is always the present.

Role Behaviour

Role behaviour is social conduct defined by reciprocal expectations (Benedict, 1969). The roles are tasks performed or fulfilled by occupants who must meet (at least minimal) expectations of how the role is performed. A case in point is the local pattern of male-care amongst the *Macaca sylvanus* in Gibraltar, where male involvement with infants is significantly greater in extent and duration than it is with conspecifics in Morocco (Burton, 1972; Deag and Crook, 1971), and where, since babysitting is a male role, young females are kept away from infants so that young males may learn their role. While the interpretation of the male-care role has certainly depended on the practitioners of it, the behaviour pattern has repeated over generations.

In Kowloon, one local group exhibits another interesting tradition of babysitting. This group has two old female Bear macaques (M. thibetana) who look after the infants of the long-tailed macaques (M. fascicularis) in their mixed group (Burton and Chan, 1987). Long-tail type females, therefore, do not babysit. The task is left to the old ones, because the expectation of that role is clearly delineated.

Because roles are conventionalized expectations of traditional behaviour, manipulation of roles for political outcomes occurs. Clearly, manipulation implies intent: a conscious use of information in order to obtain a goal. There is now ample recognition that such behaviour does occur in non-human primates (e.g., Byrne and Whiten, 1988).

> Frances is fully mature now; at her prime. She is perhaps the doyenne of the females in her group. She moves easily between individuals and sub-groups; is groomed by almost everyone. She readily approaches Sam, as doddering and crochety as he appears. Some days ago, four young males came - invaded - from the other group. They killed Jim in a fight. He lost his footing and tumbled down the cliff, crushing his skull from the impact. Amongst the four young males was Jake. Still nervous, he was noticeably tense, ready to jump or run at any moment. Frances spent some time quietly approaching Jake. She nattered, and flopped down inviting a groom. He responded, tentatively, with quick grooming strokes and not for long. The situation round and about required too much attention. But Frances was persistent. For days she kept approaching him, grooming him, asking to be groomed. When her perseverance earned his trust, she began to approach Sam and groomed him. She would then go to Jake and groom him, and alternating between one and the other, she was able to decrease the distance between them, bringing them within a couple of meters, until finally she left them an arm's length apart. Her mediation worked: Sam allowed Jake to groom him, and Jake became the new second male. (M. sylvanus)

The determination of which individual occupies what role is apparently a local phenomenon, with age-based decisions more prevalent than sex-based ones (Burton, 1977). Outside of the biological roles of progenitor and progenitrix, the basic social tasks, rearing the young, moving the troop, protecting the troop and maintaining group cohesion, could be filled by either sex (Burton, 1977).

Learning established patterns is a complex process. As the monkeys mature, their observation of what other animals do within the group establishes traditional norms or conformities, and defines roles. The young enact what they have observed. The performance of the action is refined by older monkeys who discipline the younger, channeling, modelling, constraining their behaviour so that it comes closer to resembling accustomed patterns. Two factors effect learning a role: [1] the size of the group and [2] personality. Access to information and therefore acquisition of it, depends on its visibility. The larger the number of animals in the group, the more an individual's view will be a small sample of the total. In the two local groups of barbary ape in Gibraltar, each individual could know each other personally, whereas at a primate centre holding 300 conspecifics in France, some individuals, would probably never be observed by others. The expression of their role, then, would not be available for copy by some individuals. The size of a group constrains the number of roles a given individual may have. The larger the group, the greater the role specification - protector may be a different role from leader, etc. - and the more copies a particular role will have. In a large group, the number and kinds of role models increase. In such circumstances, the choice of model(s) to imitate is hard to determine. What makes the player of a role visible may be his/her relationship to the viewer, his/her status in the group or sub-group, his/her age, or more ineffable individual attributes, and even chance.

Personality

"Personality" (Poirier, 1972) in non-human primates is a composite of development, genetics, physiology, experience, awareness and, simply, unknowables. An individual then, is a singular phenomenon. The understanding or "reading" and consequently, the expression that individual will give to a role will be unique. That interpretation, however, cannot overstep local norms, because there is social sanction: others will behave negatively or not respond. Some individuals within a group, for reasons of personality alone, have more prominence than others. Their actions are recorded whatever their quality.

> Sweetpea is running pell-mell through the group. The juveniles and infants scatter before her. She runs up to her mother, who gently threatens this juvenile, and sits down, catching her breath. The giant of a female bear macaque (*M. thibetana),* who is her mother, sits eating quietly. Several adult animals approach, and the mother threatens them all, with a minimum of energy in her gesture: she barely opens her mouth or even lifts her brows. They startle and move back. Sweetpea leaves her mother, and rushes up the hillslope behind, her goal a young juvenile who is swinging on a liana. Sweetpea grabs the little monkey's foot, and bends over to mouth it, till the juvenile, losing its balance on the swaying liana, squeals. The warning 'runnh' sound from several animals around her, sends Sweetpea dashing back to the ground near her mother. This formidable female raises her head, and glances about, extending her protection to her devil-may-care offspring. There Sweetpea attempts to tussle with a few juveniles, but unwilling to play with her, they scamper off. A human with a young child passes by, and Sweetpea, her mouth agape, her forehead pleated in threat, rushes to the child, grabbing at its clothing. Sweetpea's mother stands to face the human, once again according her recalcitrant offspring the protection of her bulk.

> Mark is old - somewhere past 20 chronological years. His incisors are gone; his brow overgrown. His shuffling walk suggests arthritic joints. Sometimes it seems his hearing (or his judgement?) is not so keen: the younger male Barbary Apes (*M. sylvanus)* prod him, cue him into reacting. A sub-adult misbehaves, chasing a juvenile, grabbing and mouthing him. The juvenile vocalizes in distress and Mark looks over in that direction. The sub-adult catches Mark's gaze and as if struck in punishment, he releases the juvenile and dashes off. Over in the other group is Sam. He is every bit as old as Mark and just as aged. A sub-adult crosses the area in front of Sam. He is chasing a juvenile, and catches him. He mouthes it, and the juvenile utters a distress vocalization. Sam looks at the sub-adult, who continues to distress the juvenile. Sam emphatically threatens the sub-adult, with mouth and eyebrow statements. The sub-adult carries on. Sam leans forward, his shoulders bobbing, the sub-adult continues to ignore him. Finally, Sam lunges towards the sub-adult, who drops the juvenile and flees.

The contribution the individual makes to the definition of the role, the personal interpretation, comes to be part of the repertoire - the behavioral options available to the members of the local group.

Because the adult retains its knowledge of its play patterns, its discoveries and what it has observed, the society is ensured a large repertoire of motor movements, social interactions, gestures, and the like. Because a primate society is age-graded, information is dynamic. There is a mechanism to retain accepted traditions, as well as adding new material. Accepted traditions are retained in the performance and modelling of behaviours by older individuals; new elements (alterations) and new patterns (innovations) are introduced largely due to the activities of the young.

The smaller the group, the greater the impact of new ways of doing things; the more likely they are to be noticed, conventionalized and included into the traditions of the local group. Inclusion occurs synchronically, through observation and imitation of one generation by the other, or diachronically, as the inventors age and remember. The number of animals in the group influences the visibility of any one of its members. Carpenter early recognized this effect of number on relationships within a group when he used the simple formulation: N(N-1)/2 to show how group number determines the number of dyads, hence the combinations of relationships. The number of roles and the number of players for the role and the representation or image of how the role is to be played, will be more variable with increasing group size.

Sperber has noted that "Representations ... tend to be transformed each time they are transmitted" (1985: 75). Individuality or personality in non-human primates is a concatenation in development of genetics, experience, perception, choice and ineffables. What image a given animal picks up from the representation of a particular role enacted by the incumbent, is unpredictable. The manner, therefore in which that individual will itself perform the role is idiosyncratic within the bounds of local tradition. Indeed, the very act of storing the information about the role, and recalling its representation alters it (Sperber, 1985). Of course, in the re-creation of the behaviour, should the animal exceed permissible norms, "retribution" will come in the form of not being responded to or incurring negative response. But within the range of intelligibility, is the space that accounts for non-biological change within the patterns of behaviour.

Variability is a function of the number of members in the group, whose several interpretations slightly alter the conventions.

Innovation

Changes in new patterns may be merely alterations or modifications in already existing forms - slight adjustments, or interpretations. Changes, on the other hand, can also be *wholly* new, that is, a novel association of elements. The term "innovation" has been used in the relevant literature to convey both meanings, unfortunately, too often without clarification. However, the consequences of *innovation* are different to those of *alteration*. The acceptance, for example, of a wholly new pattern will be different from modifications in an established one. The meaning of the term 'innovation' is therefore restricted in this chapter to refer to the introduction or creation of a totally new *pattern*. Changes in elements of an existing pattern or tradition, are alterations or, simply, modifications.

The analogy between biological and social transmission of information has been drawn above, in order to highlight the fact that information transmission is conservative and radical at the same time. In social behaviour this is a function of the generational structure of a non-human primate society. There are "mitotic" and "meiotic" parallels: replication by the older generation of stored behaviour patterns represents conservation of information. It is usually the younger generation (e.g., Jolly, 1985) which introduces new information, with innovation analogous to mutations and meiotic intrachromosomal recombinations, that is, crossovers, which realign genetic information. The juxtaposition of established elements creates a new statement. Modifications are analogous to simple meiotic interchromosomal recombination, i.e., Mendelian independent assortment: the association of parental chromosomes.

The analogy underlines the regularity of this process: recombination regularly (but not invariably) happens in meiosis; replication regularly happens in mitosis. The generational effect in a society also gives that regularity. Following the analogy with meiosis, behavioural innovations are most often due to associations of already existing information. Mutation is rare and not often incorporated into the information-set whether as genes or traditions. The biological analogy is limited, however, in that young animals, of course, can faithfully replicate behaviour patterns and older animals can innovate.

When it is older individuals that innovate, it is often due to ecological pressures (Hauser, 1988). A recent example of the development of an innovation amongst mature animals comes from the "Pump House Gang" of baboons *(Papio anubis)* on the Kekopey Ranch, near Gilgil, Kenya which prey on a variety of animals. Changes in predatory techniques have been documented since the group was first studied (Strum, 1987; 1981). After three years of predation on accidentally found animals, a new pattern of ".... elaborate searching, stalking and chasing ... complex hunting" (Strum, 1981: 274) developed from the previous casual stalking and capturing. Then the pattern of a single animal hunting gave way to coordinated efforts: ".... baboon males chased Thomson's gazelles toward other hunters, and several kills resulted from sequential chases" (Strum, 1981: 275). The proximity of infants to females who were interested in eating meat gave an impetus to the dissemination of this tradition. "Schooled," as it were, at their mother's side, their "...observation, investigation, and imitation resulted in meat-eating for infants of all ages" (Strum, 1981: 280). The relationship of infant to adult male also influenced infant participation, and hence potential absorption of this pattern.

> ...on one occasion several infants attended a kill with a male with whom they associated regularly. They had no special relationship with the male feeding on the carcass at the time and did not approach it or him. For over 45 minutes they stayed near their male associate, and when the previous consumer finally left and the second male took the carcass, the infants immediately rushed toward him and the meat, investigating the carcass and feeding on scraps that were nearby. (Strum, 1981: 280)

Eventually, the male baboons began to actually share the meat from the kill with females, moving aside to let a friend have a bit (Strum, 1987).

The first, and classic example of this process of innovation and tradition development, was observed several decades ago on Koshima Island, Japan, in a provisioned troop of *Macaca fuscata*. A 16-month-old female departed from the usual behaviour of rubbing sand off potatoes provided to the troop on the beach. She began washing the potatoes in the sea prior to eating them. This innovation, over time, spread throughout the troop. After four years, half the troop were regularly washing their potatoes, and by the end of nine years, 71% of the troop were habitually doing this (Frisch, 1968). The route by which this innovation was disseminated is interesting. The inventors, the young, were the first to imitate the behaviour which then spread to the older ones. Speed of assimilation of an innovation depends on who innovates. Some individuals, because of their personality, age, role, status or affiliations within the group will be more noticeable than others. Hence, their behaviours are more likely to be observed and imitated. The significance of the contribution of such noticeable members to the group is that their ways of doing things over time may become essential to the group.

Note that it took four years for half the troop to regularly use the potato-washing innovation. This is not only because it takes time for imitation of an imitation to diffuse through a group, but because another process is also involved. As monkeys age, they take with them into adulthood the knowledge (innovations included) they learned as young. Hence, even without imitation by one animal of another, that which was radical, and perhaps even unacceptable to the older generation becomes part of the repertoire of the troop as the young become old. The example shows how new elements are incorporated into the catalog of behaviours over time. Thus change is part of the system, and variability is the corollary or consequence. Innovations are learned as part of the group's conformities, and the absorption of these construct local ways of doing things: tradition.

Tradition

The argument over whether or not non-human primate societies have traditions has tended toward the affirmative. After a decade of evaluation of deme-restricted patterns in monkeys and apes alike (Kawai, 1965; McGrew and Tutin, 1978; McGrew, Tutin and Baldwin, 1979; Goodall, 1986; Strum, 1975), it seems that: "The capacity to acquire a given technique through learning plays a role similar to that played by genetically controlled behaviour in lower species in ensuring the continuation of some behaviors..." (Goodall, 1986: 562).

Traditions are patterns of behaviour that are endemic to a local group and that are acquired by social transmission from one individual to another and from one generation to another. Many traditions that have been witnessed concern food acquisition, as with the Japanese macaques (Kawai, 1965), or termite hunting amongst chimpanzees (Goodall, 1986), or hunting behaviour amongst baboons (Strum, 1975). Some observations have been made on tool-use amongst apes (McGrew and Tutin, 1978; McGrew, Tutin and Baldwin, 1979) or the development of social traditions, as in male-care in barbary macaques (Burton, 1972; Burton and Bick, 1972). The actual number of witnessed, confirmed, and fully described traditions remains small and has limited the discussion of Tradition in non-human primate studies. It may well be that this limitation is due to the inability of the observer to witness tradition.

Traditions are the conformities or social dictates which members of a group obey (Hall, 1968). Amongst these are when and where to play; which individual is available as sexual partner; whom to sit next to; what food is edible; where food may be found at certain parts of the year; who raises the infants, and so on, so that the roles that individuals occupy are defined by social norms (Burton, 1977). A group's traditions defines the set of conventions by which it operates including foraging, socialization, hierarchies and role behaviour.

Traditions are conventionalized representations. Built from reactions, and reactions to reactions within the local group, these conventionalized representations, over time, constitute that group's tradition. Representations are information and a non-human primate society is a dynamic system of information. It is dynamic because in non-human primate societies the delicate balance between stability and instability is a permanent state. The tension between the conservative forces of replication, of reduplication of tried-and-tested patterns which are stable, is in dynamic opposition or synergism with the forces of mutation, of radical change, which are unstable. The oldest members of the group stabilize information, while the young, particularly through their patterns of play, introduce new elements. Because typical primate groups are comprised of three generations, the conservation of pattern is balanced with the introduction of new elements. As long as the oldest member of the group can retrieve a stored image, that representation can potentially be conventionalized and inducted into the group's tradition. Part of the social value of the old in a primate society lies in this storage capacity (Rowell, 1966). Remembered in turn by the youngest member of the group, traditions can have considerable time depth.

In his discussion of local traditions and "cultural" transmission in non-human primates, Nishida suggests that social knowledge is "lost when the individual dies" (1987: 463). But he disregards the fact that however transient a behaviour, its performance, its action time leaves an imprint on whomever has observed it or knows about it. As these are remembered, the time factor extends to the duration of that storage system - i.e., until the last animal who witnessed the last performance of that action dies. The action-memory that is the result of behaviour, then, has a reality because it potentially continues to effect others. It is the source of social networks, hierarchies, patterns of foraging, nest building, socializing, aggression. To that extent, it is *transcendent* because the effect goes beyond the individual creating it. The behaviour or sum of behaviours, while a function of organic being, muscles, bones, hormones and cells, exists beyond the organism that behaved within the memories of those who witnessed it. The set of conventions which is the weave of the social fabric is unique to a local group.

Manipulation of Representation

Shared representations constitute a set of information local to a particular group. Manipulation of shared communicative elements occurs, and indicates the ability to utilize representations for purposive or specific goals (Burton, 1984; Cheney and Seyfarth, 1985). Manipulation of representations depends on socially agreed upon representations, that is, forms of communication and behaviour that are responded to in regular or ritualized ways. These constitute the conventions of a group. Socialization is the intergenerational transmission of culture (Williams, 1975), where "culture" is understood as group local "information" (McFeat, 1974). Socialization is the process of maturation within a group such that witness of, and ultimately replication of conventions ensure information transfer, with relatively little noise or distortion in daily circumstances. The more clearly an individual knows patterns, the greater the guarantee that information will be transmitted as sent. Socialization is the process whereby group local patterns or conventions are assimilated. By altering conventions, an individual can achieve a goal. Manipulation of one's conventionalized representation is commonly done. Typically, an individual employs gestures belonging to another age/sex category. A particular response to this gesture is expected. Utilization of an inappropriate representation itself, therefore, represents a sophisticated awareness of cause and effect based on prediction of outcome (Gouzoules, Gouzoules and Marler, 1984).

This is particularly so if an individual utilizes an infantile gesture because these are *obligate* stimuli.

> I am watching three *Cercopithecus neglectus* monkeys this morning. The adult female, Bella, is chewing some delectable bit of banana; her juvenile, Beebee is looking around. The adult male is sitting, doing nothing in particular. Beebee's gaze wanders rather directedly towards the male, her father, and then to me; to her mother, then to me. Suddenly, she shrieks, and I am surprised that she uses an infant's shriek. The male rushes at me, his threat is frightening enough - even with bars - that I back off.

This powerful stimulus apparently obliged the adult male to respond by charging the intruder. The juvenile employed a conventionalized representation for a goal; the adult male conformed to a public representation in fulfilling his role. This anecdote appears to support Sperber's (1985) contention that other animals cannot disbelieve what they perceive or decode, because the adult responded to what *appeared* to be an infantile cry. Other information however, suggests that what occurred here is considerably more complicated: the adult male apparently chose to respond to the juvenile as if the communication was *bona fide.*

Surely, having lived with the juvenile for some time, and knowing the noises that infants make, he will have known that what the juvenile was uttering wasn't "true." The ability to manipulate information with strategic intent, that is, to deceive, is documented under experimental conditions (Byrne and Whiten, 1988; Goodall, 1986; Woodruff and Premack, 1979; Menzel, 1971). Success at social manipulation depends on knowledge of and "insight into the social networks of a group" (Strum, 1987: 121). What this monkey anecdote suggests however, goes one step further: the receiver evaluates the representation, and evaluates the transmitter as well. Hence, the response is qualified. Manipulation is a term preferred to "Machiavellian Intelligence" as it is a neutral one which does not judge the motivation in assessing the act. The ability to manipulate a representation stems from a more general primate ability. Primates are unique in having several visual centers (Passingham, 1982), and in their ability to quickly associate images and thereby to *predict* situations. It is the ability to PREDICT which is the fundamental ability underlying primate intelligence. Manipulating images is a subset of this ability, a facet or derivation of which is the ability to deceive.

Non-human primates do make mistakes, however. A threat face directed at a particular monkey may be disbelieved, and the threatening animal approached anyway. Perhaps some quality of the gesture, some meta-quality, modified it to mean "I look as if I were threatening but am not." On the other hand, it may also be that the receiver simply misinterpreted, or chose to appear to misinterpret, and approached anyway. If the sender then attacked, we would know the threat face was intended and the receiver had misjudged: the act of approaching, however, may alter the behaviour of the sender, confounding interpretation. The classic example at this level, is the non-threat yawn. The primate clarifies the representation by covering the canines with the upper lip: it "reads" therefore, "disbelieve that I am threatening." One part of an expression, such as the eye component, can belie what another part, such as the mouth component is stating (Zeller, 1980). Thus, a macaque female can threaten with the eyes but lip-smack simultaneously. The viewer receives a single message which is understood as a meta-representation precisely because it is ambiguous. The doubt cast upon the image is derived from the juxtaposition of incompatible elements. Repetition of this juxtaposition conventionalizes it.

Tradition Drift

The traditions of a group, then, are not static. There is a drift in a group's traditions as patterns diverge from each other over time or space: group A is different to its descendant group A, or its fissioned group B. Tradition Drift (Burton, 1972; Burton and Bick, 1972) comes about for a variety of factors: [1] there is variation in the production of conventions; [2] innovation occurs; [3] the presentation of representations by some individuals by virtue of age, sex or status, are prominent and are observed; [4] the caprice of memory stores some representations and not others.

The concept of Tradition Drift contrasts with what Eggan (1963) and others before him (e.g., Herskovits, 1948) termed cultural drift. This idea was presented as an adaptationist explanation and described as "....a regular series of changes in social, political, economic and religious institutions, a series which has a definite direction" (Eggan, 1963: 253) and this direction of change was due to "....selection as a result of internal adjustment of the interests of various groups within the society" (Op. cit.: 254).

Tradition Drift is the non-biological parallel to the genetic process: an alteration in the pattern of information which is not due to Natural Selection (Suzuki et al., 1989). In this, the notion contrasts with Wilson's use of it:

> The component of divergence based *purely* on differences in *experience* can be referred to as tradition drift (Burton, 1972). The amount of variance within a population of societies is the sum of the variances due to genetic drift, tradition drift, and their interaction. (Wilson, 1975: 13-14) [italics mine]

That is, agreeing that Tradition Drift accounts for differences between groups, Wilson goes on to relegate those differences to random sampling error of the *genetic* endowment of individuals which, he argues, determines social behaviour.

> In any particular case the genetic and tradition components will be difficult to tease apart and to measure. Even if the alteration in social structure of a group is due to a behavioral change in a single key individual, we cannot be sure that this member was not predisposed to the act by a distinctive capability or temperament conferred by a particular set of genes. (Ibid: 14)

Certainly, the act or behaviour must stem from its biological basis, must work within its morphological constraints, but the act and the consequences and products of the act exist in their own right. Behaviour, as Piaget succinctly put it, is the "motor of evolution" (Piaget, 1978: 146) - not its consequence. Tradition Drift is a *social* process.

Biologically, all individuals in a group are not equally relevant, since biological transmission of a non-human primate group is effected by a small number of individuals. Whatever the original or absolute size of the group, the Effective Population Size is a much smaller fraction. Biologically, the probability of any member of the group being expressed in the next generation is a function of the number of genes it shares with the actual reproducers. This means that the representation of the group depends on a few.

For the social system, however, the individual has a different meaning since all associations, that is transfers of information, are equally relevant, despite the fact that not all individuals in a non-human primate group are equally visible. Hierarchies of age, gender or status, do mean that some members of a group are more prominent. Hence, as the Japanese experiments with the diffusion of new foods in a troop showed more than twenty years ago (Itani and Nishimura, 1973), innovation by a prominent individual spreads rapidly through a group. Innovation, however, regularly spreads because the innovator is in close association with relatives and peers and others who imitate its behaviour (Fagen, 1981). In this fashion, young individuals can be the source of the diffusion of novel information. It is not their prominence, but their associations which make them relevant. Whereas, however, the number of fertile matings is a reducing number, the number of associations is multiplicative (as witness Carpenter's formula for dyads above). Hence, in terms of social impact, all associations, that is transfers of information, are equally relevant and each individual is significant to its group.

As a metaphor for a non-biological mechanism, Tradition Drift explains differences between groups. The metaphor is helpful in evoking the notion of generational fluctuations with no preordained direction; no intent or purpose. Traditions change, not because it is good, bad or indifferent or even adaptive for them to do so, but due to the vagaries of the process of absorption and representation. As with genetic drift, change occurs dramatically in small groups, though given sufficient time, in any group (Suzuki et al., 1989). The metaphor ends there, however, because genetic processes involve *traits* subject to statistical predictabilities, while behavioural traditions are *patterns* - larger concatenations of integrated wholes, some of which allude the researcher's measurement. Then, too, Genetic Drift is truly a random process, that is, a statistical phenomenon where any item amongst several has precisely the same probability of being included in any sample of N observations (Hays, 1973).

The factors effecting temporal continuity of a tradition violate this assumption. They include visibility, personality, affiliation, status, and age. Yet, there is no, and indeed there can be no evidence, either, that these components are "selected," that is, that a pattern as expressed by an individual has a direct influence on the number of offspring's offspring that survive. The influence of an aged personality, like Mark in Gibraltar, or Fatty in Kowloon, is undiminished despite his/her attenuated reproductive role. On the other hand, the genetic value, the future adaptive value of e.g. washing sweet potatoes as innovated by an older juvenile is, to say the least, obscure. That is because it is a proximate behaviour.

For some biologists (e.g., Dunbar, 1988; Tooby and Devore, 1987) proximate behaviours are significant only in the service of evolutionary process. To those participating in them, however, proximate behaviours are the totality of existence. Having a new food source NOW ensures fullness NOW and another food source in the repertoire tomorrow. That in turn affects the number of animals within the group and the relationships between them and that series of consequences goes spiraling out and at the same time feeds back with new consequences - all of which are still proximate.

The relationship between the here and now and evolutionary process is a guess. Like religion, it becomes an article of faith, that the complex behaviours occurring in the presence of an observer, have meaning - value - only in the distant future. Renunciation of the importance and meaning of this life for some other "time" (heaven, tomorrow, evolutionary future) overlooks the impact of current activity on the nearby interval, and neglects the subject's reality. A monkey generation may be four or six years; its lifespan 20-25. That monkeys do not choose the mates for their offspring let alone their offsprings' offspring suggests that behaviour and even political machinations are about themselves and not merely promulgation of genes. Even political behaviours (de Waal, 1982; Mason, 1988) which can extend over several years' duration have their significance in that immediate future: both in the process, the machinations themselves, and in the outcome.

> Joan strokes and parts Bridget's fur. Intent on her task, her hands move almost rhythmically. Bridget, relaxed, sits with eyes closed. Bridget's daughter is nearby. Wilma, Bridget's sister, sits a considerable distance away. It's been like that since last summer, when Joan began "courting" Bridget and keeping Wilma away, challenging Wilma but soothing Bridget with grooms and natters. Wilma is approaching now. Bridget stands, and, surprisingly, so does Bridget's daughter. They face Wilma, and the daughter, watching Joan, taking her cue from Joan's face, also looks at Wilma, raises her eyebrows, drops her jaw, threatens her mother's sister so she cannot come near. Joan has been following a young male as well. She sits with him, follows him, grooms him, is the active partner with the reluctant male. He walks with her, she strolls further and further away from the main group with him in tow. Are all these intrigues, this separation of Wilma and Bridget, this reorienting of Bridget's family towards herself, this alliance with the young male, is this about forming a new troop?

Conclusion

Non-human primates live in a proximal world enmeshed in their interactions. Each member of a social group is the individual result of a long personal history. That history includes genetic templates for the manufacture of organismic substances - from bone to hormone - development in an ecological context and social community (maturation and socialization respectively) and results in the personality, the being which is, clearly, not just a summation of its parts. Group members learn the codes which constitute the repertoire of the group. While the individual is maturing, is developing, is being socialized, it is also active. That activity includes selection, imitation, manufacture, amalgamation, emphasis, innovation, and perception of representations. When individuals interact, the result is social behaviour. Regular patterns of doing things form traditions in a society. Non-human primates are capable of recognizing the *impact* of their representations and of manipulating them to some purpose.

Conventionalization of representations constitutes a local group's traditions. Over time, and through space these patterns change. Tradition Drift comes about due to a variety of factors: there is variation in the production of conventions; innovation occurs; the presentation of representations by some individuals by virtue of age, sex or status, are prominent and are observed; the caprice of memory stores some representations and not others. Tradition Drift accounts for the differences between groups in their social behaviour. In addressing the question of what is the locus of social behaviour, Tradition Drift explains the process of behavioural differentiation between generations of a society, and between neighbouring, recently fissioned demes.

Acknowledgements

A version of this paper was first presented at the International Summer Institute for Semiotic and Structural Studies at the University of British Columbia, Vancouver, August 1988, and was subsequently presented to the Social/Cultural Anthropology Colloquium, Department of Anthropology, University of Toronto, March, 1989. Comments and incisive questions from these meetings were most appreciated. I especially wish to thank Profs. Tom McFeat and J. Chew for their thoughtful reading and comments on the original manuscript.

CHAPTER THREE

Communication in the Social Unit

Anne Zeller

There are many types and levels of communication in all animals from the one-celled organism to the most successful socially complex primate. This multitude of forms communicate for many different purposes, and some use modalities such as electric charge or flashing light that are not available to others. The basic functions of this aspect of behaviour serve to promote survival and reproduction by aiding in the acquisition of mates and food, and in protection from predation. In more socially complex animals such as primates, communicative skills are expanded which allows them to establish and maintain long-term social bonds, and to socialize their young. Due to the life span of the animals and the complexity of their social relationships, the development of a social unit among primates is a lengthy process, fraught with tension and potential misunderstandings as observation of newly formed groups frequently demonstrates (Bernstein, 1971; Bernstein et al., 1974). The maintenance of group bonds and the process of socializing newcomers, whether offspring or strangers, as well as the exigencies of everyday life, is one which requires a complex and well-understood system of transmitting information.

In any interchange between individuals there are always a number of options available to the interactants. In particular there is always the choice to respond or not to respond to a message being sent. Cases have been recorded in which receivers are observed to look away or close their eyes to avoid receiving messages from other primates (Fedigan, 1982). The sender also has the choice of whether or not to transmit a message, and the judgement of timing in both action and persistence will reflect the intensity of the need to communicate. The outcome of the interaction can also be radically altered if third parties are involved. A young animal may be able to evoke a particular reaction from adults when in close proximity to its mother, but receive quite a different response to the same message in her absence (e.g., Bramblett, 1976; Burton, 1984; Ransom and Rowell, 1972). These are only a few examples of the complex influences affecting the operations of primate communication systems. The present discussion is intended to examine the function of such systems in establishing and maintaining the integration of social groups. In order to fulfil this aim, some data on the structural aspects of communication is presented, with particular emphasis on the relationship between structure and function. The problems of understanding this system of behaviour are further magnified by the lack of agreement about the range of behaviours that can be legitimately included under this heading. I have chosen to define the term broadly because all of the information transmitted from one individual to another increases the predictability of behaviour and thus the potential for social understanding.

When observing the interactions of a group of primates, it becomes clear that information is transmitted in a wide variety of ways. Because of the importance of understanding the motivations of others and reaping information from their actions it has been said that everything social that a primate does can be classified as communication since it transmits information (Altmann, 1967). At one end of the spectrum an understanding of the process can be approached by listing the channels or sensory pathways through which information can be sent and received (Fig. 1).

Fig. 1 - Channels and Modalities of Communication

Channels		Modalities	
visual	facial gestures	proxemics	kinesics
auditory	vocal	non-vocal	
olfactory	pheremones	scent marking	other
tactile	grooming	body tension	touching

There are four major channels used by primates; Visual, Auditory, Tactile and Olfactory. Some of these channels, such as the Visual, process information that is coded in a number of different ways which are generally called modalities. For example, the visual channel allows the coding of facial expression, hand gestures, body position, and the spatial position of the animal in relation to others or its environment. Body position and spatial relationships are called kinesics and proxemics respectively. Sources of data which are coded in these modalities include movement, colour, position of body parts and level of tension, which interact to provide considerable subtlety in the nature of the message sent by that channel alone.

However, this listing of channels and modalities, while providing some scope for the understanding of the various ways by which primates transmit information, does not take into account the multiplicity of channels used concurrently (van Hooff, 1967), the level of volition involved in sending the message (Menzel, 1979), skill at coding (Struhsaker, 1967), attempts to suppress the message (de Waal, 1982, 1989), the sending of false data (Woodruff and Premack, 1979) and the level of success that the sender has at obtaining the desired response (Miller, 1967, 1972, 1975). The situation is further elaborated by the probability that the interaction is not truly dyadic, but influenced by third parties to varying extents (Reynolds, 1970; Burton, chapt. 2). Therefore, in addition to considering the modality and channel of a message, my approach to communication also includes a consideration of the level of volition and control over sending that individuals possess.

There is a whole set of information about the individual that cannot be hidden and does not require any intentionality to send. I have called this the 'basic' level of communication. The individual's sex, age, identity, size, and species are all evident through the visual channel to any conspecific looking at the sender. Much of this data is also evident through the olfactory and auditory channels, especially if the two animals are known to each other. For females of many species the state of the estrus cycle is evident in the visual and olfactory channels. An individual's general level of health can also be transmitted through visual and olfactory cues as well as any distress vocalizations it may make. Thus a considerable amount of data is available about another animal before any interaction occurs. The second level of communication I have called the 'interactive' level and attribute some choice to the individual about what kinds and how much information is transmitted. The four channels and many of the modalities listed in Figure 1 can disseminate information at both the basic and interactive levels. Each level of data reinforces the other. The message is X (data) but it is being sent (interactive) by individual Y (basic) with status Z (social) which means that not responding is more likely to have serious consequences than if message X was sent by another animal. The nature of the message, its intensity, and the individual it is directed towards are all matters of choice on the part of the sender, but once the message is sent, the identity of the sender is known and response choices can be made based on previous encounters. The importance of this relationship between various information levels is one reason why the system works more easily and predictably with known animals than with strangers.

Functions of Communication

The social unit, therefore, is the setting and focus for most of the communicating that occurs among group-living primates. Information is transmitted throughout the group at many levels. Three areas of social relevance which will be discussed in this paper are messages of group-wide importance, the establishment of long-term social relationships, and the immediate transmissions which are the foundation of day-to-day information flow. These three functions were chosen for discussion because they reflect, in progressing order, information transfer to large numbers of individuals, and inter-individual exchanges. All of these uses are dependent on factors underlying the functioning of the communication system. These include the age and social position of the senders, their level of influence through rank and kin relations, their ability to code information so that others can easily comprehend it and the willingness of others to respond to them.

Group Messages

One of the important functions of group-wide messages is the maintenance of social cohesiveness. Animals of a group usually stay within visual contact with the others, especially on the ground. However, arboreal animals frequently need to utilize contact calls to maintain group cohesion. Sometimes if an animal is lost, it may use loud contact calls in an effort to relocate the group, but this is a subset of the general function of maintaining group cohesion. A major group-integrating social phenomenon is the existence of social roles which are a number of categories of behaviour necessary for the maintenance of the group but which do not have to be undertaken by all animals (Reynolds, 1970; Burton, 1977). The role of sentry is a very important one, because it allows most of the animals to concentrate on foraging, or to rest comfortably, while only a few keep a look out. Social factors, however, influence even this seemingly straightforward task since an alarm bark by one animal may be responded to much more quickly than an alarm bark by another, even though both may be equally valid (Rowell, 1972). Another role with communicative function is performed by individuals who initiate the direction and speed of group movements, as well as by the possibly different animals who consolidate and ratify such decisions by their participation (Burton 1972, 1977; Kummer, 1968). Burton (1972) states that the head male of a Barbary macaque group can either initiate group movements or ratify them by following an adult female who has moved away. In some cases, an adult female will refuse to follow the group and thus limit troop movement because the male will return to her. I have seen a male macaque in such a situation try to drag a reluctant female to her feet to follow the group, but in vain. In other species the indication is more subtle and there may be several false starts as different directions are indicated by various animals until the major part of the troop follows one of the initiators. Once a direction has been established, it is important for the group members to move with the others since those who lag behind or go a different way are in serious danger of not being seen again due to predation. This level of cohesion is less important for arboreal forms with small ranges, but even they tend to stay in communicative contact because of the safety provided by animals filling the sentry role. The potential for high-ranking individuals to intervene in the agonistic interactions of others to prevent the escalation of fighting is another community-wide influence. This is frequently called control behaviour or a 'control role' (Bernstein, 1966; Bernstein and Sharp, 1966), and may be carried out by merely a look, by threats, or by actual attack depending on the level of social influence wielded by the control animal. The recognition and maintenance of group boundaries and identities is an aspect of group information flow which still poses a number of research questions since primate groups are not all structured in the same way. Pair-bonded animals such as gibbons live with their immature offspring and are not usually open to immigration by other conspecifics. Their response to non-group animals is to approach, call, chase across group boundaries, and occasionally fight with conspecifics attempting to encroach on their territories. Occasionally single immature animals can become associated with a gibbon group (Quiatt, 1987), but as offspring mature they tend to be harassed by the resident adults and eventually they move out on their own.

A different pattern of social organization which clearly emphasizes the 'we/they' dichotomy is seen in animals with less cohesive group structures such as the chimpanzee. In this pattern, the animals who form one community are not frequently all together in one place. When parties of animals from the same group meet up there are often extensive greeting rituals, including touching, presenting, mounting, charging, hooting and drumming displays. Some of these can be quite energetic but are not aggressive although young animals can be hurt if they do not stay out of the way. Chimpanzee responses to strangers are quite different. Most of the migration among chimpanzees occurs when females move from group to group. Normally they do this while in estrus and although resident females will attack them, the resident males of the new group will usually offer them some protection (Goodall 1971). Previously resident females can return to their natal groups although not many clearcut cases have been observed. Non-estrous females and male strangers who move into another chimpanzee range are frequently severely attacked, jumped on, pounded, dragged and bitten and sometimes die from these encounters (Goodall 1986).

> In the first instance, five males ... suddenly raced forward, there was an outburst of screaming and barking, and when Bygott (the observer) caught up, he found some males attacking, while others displayed around an unhabituated female. Finally she escaped; Humphrey next appeared holding an infant chimpanzee, between 2 and 3 years old, which he and Mike killed by eating.... A few days later, Humphrey, Joma, Figgan and Faben were again observed as they chased and severely attacked an unhabituated female thought by Bygott, to be the same individual. (Goodall, et al., 1979: 32)

In describing an attack on a Chimpanzee from another group, Goodall said:

> Jomeo instantly leapt down, seized De's leg and dragged him to the ground. Evered now joined the attack: De lay flat on the ground and no longer tried to escape as the three males assaulted him, all screaming. Little Bee continued to watch and Gigi began a display, stamping and slapping the ground, and then joining in the attack. The fight became even more vicious, as all four Kasakela chimpanzees hit and stamped again and again on the still prostrate De who was uttering a few squeaks. The observers saw the aggressors tear skin from De's legs with their teeth, and at some point he was dragged along the ground. (Goodall et al., 1979: 37)

The attacked animal, De, was seen in very poor condition two months later, but subsequently disappeared (Goodall et al., 1979). Among gorillas, the movement of young males into a group is sometimes tolerated, but mature males are usually attacked by resident silverbacks and sometimes other group members are hurt in the fighting (Fossey, 1983; Schaller, 1965). A common pattern of male migration is seen in a large number of primate groups with either a single-or multi-male focus. In these cases many males undergo the process of working their way into a new group and most groups have some immigrant members. Known animals are responded to in quite a different way from strangers. The newcomers often spend periods of time at the periphery of the new troop subject to periodic harassment and chasing if they come too close. Even in species which permit immigration from other groups, the procedure of changing from an out-group individual to an accepted one requires communicative skills to develop positive social relations with the members a new group. Various strategies are used to facilitate this change. Burton (1984) described a male Barbary macaque who engaged in a complex communicative interaction with a female who was a resident of the group he was trying to enter. When she approached to present and groom, he avoided her advances "but did so while still making the approach face: the toothchatter towards her" (Burton, 1984: 74). When she tried to groom him "[he] threatened her mildly with his eyes while tooth chattering with his mouth" (Burton, 1984: 74). The male did work his way into the group with the aid of this female who managed to adapt to these contradictory signals. In some cases, animals already know others in the group, perhaps previous migrants, and can interact as known animals with them. This may improve their confidence and encourage others to accept them more readily. Another strategy is to develop positive social relations with a female, by attempting to groom, consort, or by interacting with her offspring, and use this friendship as a base for establishing other positive social relations. As has been observed in baboon groups, some animals do not seem to have the skills required to establish a place in a new group and are not accepted. After a period of harassment (chasing, tail pulling and biting) or fear responses and ostracization, they frequently leave and try another group (Altmann and Altmann, 1970; Smuts, 1985). Smuts (1985) observed many successful immigrations into the baboon troops she was studying and several, such as Ian's, which seemed to fail through a lack of communication skills.

> Ian, (an incoming migrant) however seemed to lack entirely the social skills necessary to placate nervous females. When a female showed alarm, he would continue his approach, and when she turned to flee, he would chase her. Sometimes the female would scream, and this tended to evoke a mobbing response from nearby animals. I saw Ian chased out of the troop by a horde of females and juveniles on several occasions.... His relationships with EC males did not cause Ian's peripheral status; he was larger than Handel and more intimidating, but his extreme lack of popularity with females kept him from becoming fully integrated into the troop. He disappeared in August 1978 and probably transferred to another troop. (Smuts, 1985: 211)

Most baboon males when approaching females under these circumstances would respond to the fear grins and gecks which indicate alarm on the female's part by ceasing their approach, sitting and making reassuring facial gestures such as lipsmacks and soft grunts (Smuts, 1985). Handel, the other male mentioned in the example, utilized his communicative skills more expertly, and was integrated into the EC group (Smuts, 1985).

Individual Relations

The development of long-term relationships is the result of a considerable amount of communicative endeavour. Animals must interact with others to develop and maintain the level of influence they hold in the group. Kin relations are expressed through communicative interactions, and their nature affects all of the personal bonds an animal develops. Socialization of the young involves developing awareness of and ability to manipulate group patterns and relations between and among individuals. Some animals can choose freely where and with whom they will sit, while others move away if approached by particular individuals and show varying degrees of tentativeness when approaching others. Indicators of confidence include body posture, tail position, muscle tension, eye contact, piloerection and facial expression. Sometimes challenges occur as one animal acts or interacts in such a way as to claim privileges which it did not previously hold. Usually this involves a series of confrontations between two or more animals, but can also occur uncontested (Reynolds, 1970). Changes of this type are noted by the whole troop, who over time alter their behaviour towards the interactants according to the outcome.

Important social bonds in primate groups, including the mother offspring bond, kin relations, sexual associations, and inter-individual affiliations are based on communication skills. The nature and importance of kin bonds in a species will influence the quality and intensity of the mother offspring bonds. Patterns of kin relationships affecting group structure encompass cores of closely related females, unrelated females focused on a male, pair bonding and less closely related females interacting with a core of mutually supportive males. There are also species living in a dispersed pattern called a 'noyau', in which mother offspring bonds are the main social focus.

In groups with a matrifocal basis such as baboons, macaques and vervets, the mother-offspring bond is one of the longest lasting. These are often large multimale multifemale groups with a long time depth and fairly complex kin networks. However, even among animals as solitary as orangutans, allogrooming among adults is seen much more frequently between mother/adult female offspring pairs than between other pairs of animals. In situations where kin bonds are an important factor structuring social life, the expression of support and amiability is easily recognised. Mother offspring pairs and sisters frequently sit together, groom, eat, sleep and travel together, defend each other and each other's offspring, and mutually benefit from social changes such as a rise in rank which may happen to one of them. Agonistic encounters between such pairs are rare and easily resolved. Young infants of a sister or mother may be carried by a relation, even when the biological mother is present. The animals engaged in this caretaking may be sibs or offspring of sibs raised in close proximity, or in other cases, as in Barbary macaques, distantly or unrelated subadults (Burton, 1972). Male/female sib pairs usually spend less time in close proximity but in kin oriented groups will frequently support each other in agonistic interactions.

Depending on the species pattern, groups may have a single or multi male organization and still be structured with a core of matriline related females. This will produce quite a different socialization environment than a structure such as a hamadryas unit, *(Papio hamadryas),* in which the females' bonds are primarily with the male, rather than each other. There are several species pairs in which the relationship between female bonds and group structure are evident. Hamadryas and savanna baboons are one such species pair in which hamadryas females tend to focus their attention on the male and are not usually closely related to the other females of their unit groups. Their mother offspring bonds seem to be of less duration than those of savanna baboons, who live in female centered cohorts of related animals and generally focus their 'grooming with' and 'sitting by' attention on these females rather than the males. This behaviour/ kinship correlation is also evident when comparing chimpanzees and gorillas. In many cases, chimpanzee females associate with mothers, sisters and brothers repeatedly and over long time periods. They forage, rest and groom together and support each other in agonistic interactions with others. Gorilla females tend to focus their attention on their offspring or the male rather than on other females, and this may partly be due to the tighter group structure in which all animals forage and move together. The close spacing, eye contact, mutual grooming and interactive support among females are not as evident as they are among chimpanzee kin.

The long period of infant development in chimpanzees may influence their bonding pattern, since juveniles must share their mother's attention with new infants for several years and usually play with and help socialize their younger siblings. Since chimpanzees usually do not reproduce before age 12 in the wild, they have the opportunity to interact with several siblings before they have offspring of their own. Positive social affiliations continue after adolescents mature, and both male and female adult offspring associate with each other and their mothers. Gorillas tend to mature a little more quickly than do chimpanzees, and while females sometimes remain in the same groups as their mothers, they frequently move out to form new groups with outside males (Fossey, 1983). Although many individuals are involved in socializing young members into the group, mothers tend to play an important role in this regard:

> ...but it was through the female that the infant was imprinted to the troop. Perhaps because of its biological need of her as a young animal, and through the grooming association thereafter, it is through her that the connection with the troop is predominantly reinforced (Burton 1972: 53).

A matrilineal organization usually provides the mother with some alternate care takers in the form of juvenile offspring or related females who will watch and play with their young for varying time periods. This allomothering behaviour is particularly noticeable among several langur species *(Presbytis* sp.), Japanese macaques *(M. fuscata),* savanna baboons, *(Papio* sp.), and vervets *(Cercopithecus aethiops),* as well as being present in a number of other species (Burton, 1977). Thus, the infants are exposed to other older group members from quite a young age and must learn to interact with them. Usually a very young animal has only a small communicative repertoire, but it must be able to signal hunger and distress in order to attract either its caretaker's or its mother's attention when necessary. The unanswered distress vocalizations of young can elicit a particular call from other adult female macaques, which I have heard continue until the infant is retrieved and quieted. Young primates also quickly learn the social approach face, called lip-smacking, which is responded to by positive approach from adults. This is one of the first gestures they are exposed to, since others approaching a mother and new infant display frequent and exaggerated bouts of lip-smacking (Burton, 1972). Infants are also genitally inspected by adults, who then often hold or groom them. Thus the infant develops positive social relations within the group outside the mother-offspring focus and this may aid it if by mischance it becomes orphaned. If young orphaned infants are adopted by non kin, it is frequently by animals who have previously interacted with them. Infants also have the opportunity to observe the interactions of other animals. As they develop they frequently establish play groups with others of similar age, usually under the watchful eye of an older animal (Burton, 1972; Rowell, 1972). By this stage they can communicate a range of messages and have observed their mothers interacting with a variety of other individuals. During their early lives while they still have their neonatal coat colour which marks them as young infants, they are not usually punished for not responding to the gestures of mature animals (Altmann, 1980; Booth, 1962). As they grow older and acquire the species coat, the adults become more demanding of an appropriate response and the youngster quickly begins to respond appropriately to the gestural and vocal system.

In those groups which are based on a pair bond, if there is only one offspring it does not usually have an age-mate to play with. In these cases, both the mother and the father usually interact with the infant, with the male taking on a gradually increasing role as the infant matures. If an immature sibling is present it may not have access to the infant to play with until the infant is able to locomote safely on its own. Marmoset infants are frequently born as twins, and thus have a play companion their own age. Since marmosets often breed twice yearly their older siblings are often still quite juvenile and playful themselves, and spend a lot of time interacting with the infants. The males also contribute markedly to the care and socialization among marmosets and tamarins, at a younger age than is usually common among primates (Epple, 1978; Wright, 1990). For all of these young caretakers, interaction with infants provides valuable experience which contributes to their future success as parents. Learning the appropriate responses to infant signals is an important part of developing the ability to rear young animals successfully. Even brief episodes of practice with infants, by quite young animals, can have beneficial effects on their nurturant abilities years later (Epple, 1978). Thus, the ability to respond adequately to the set of cues provided by a young infant is one aspect of socialization which is as valuable as the ability to interact with peers and adults.

A whole repertoire of communication focuses around the establishment of breeding bonds among primates. In solitary or pair-bonded species, and in some single male - multifemale situations, the mating bond is the strongest and is expressed by proximity, synchrony of behaviours and mutual grooming. Because of the size of such groups, multifamily bonds are not usually developed. The actual process of establishing long-term pair bonds has not frequently been observed, especially in the wild. Among gibbons the intercalated nature of their duet call indicates that two animals are interacting in a mutually supportive way. The 'half a song' that a lone individual gives is undoubtedly a cue to another individual that a potential mate is nearby. Although gibbon pairs are long-term relationships, some animals do die of age, illness or accident and the survivors must then seek new mates. They may stay in their established territories and call, or if they are unable to maintain their territory alone, they may move in search of a mate, responding to the calls of lone animals (Deputte, 1982).

The process of establishing single-male unit groups has been more frequently studied. Among hamadryas baboons (*Papio hamadryas*), the male usually takes the initiative if an unattached female is encountered. He will often approach, touch her, eyebrows raise, attempt to groom briefly and then attempt to get the female to follow him. If the female is quite young, he may treat her more as a juvenile than a mate by carrying her and helping her negotiate the steep cliffs (Kummer, 1971).

When the female matures, she then becomes a breeding member of the unit. Females may follow closely and groom the male frequently thus indicating that they accept him as a mate, or they may keep as far away as they can and look around at other animals. In this case another male may challenge the unit leader and the female may quickly move towards the challenger into his sphere of influence (Bachmann and Kummer, 1980). In these confrontations often the unit male will try to grab and hold the female who is occasionally hurt in the process. Usually hamadryas males do not challenge each other for females unless the female indicates by spacing and lack of attention that she is not committed to following that particular male.

In species where short-term consort relationships occur, such as various macaques *(Macaca),* savanna baboons *(Papio),* mandrills *(Mandrillus),* mangabeys *(Cercocebus),* patas monkeys *(Erythrocebus),* orangutans *(Pongo),* and some prosimians, each time a female goes into estrus, she will establish mating associations. During the early phase of estrus she will often be mounted at brief intervals by a number of different males. Females may indicate an interest in a particular male by approaching him, giving mating calls, and grooming him repeatedly. Particular female proceptive behaviours, such as head flagging, rhythmic tongue protrusion, and male genital stimulation are frequently seen among langurs *(Presbytis),* Japanese macaques and orangutans (Fedigan, 1982; Galdikas, 1979). As the estrus cycle advances, more males are usually interested in the female and one may begin to consort with her, mating repeatedly, grooming, eating with, and staying beside the female, usually towards the edge of the group. Their status as a consorting pair is made evident by their close spatial association, frequent mounting, and the male's aggressive response to the approach of another male. Sometimes this communication of mating relationship results in attacks on either partner by other males or females in the group. There are females who frequently attempt to interfere in a consort relationship by attacking the consorting female, and I have seen a very old female mandrill whose aggressive behaviour effectively terminated another female's mating relationship. Males also genitally inspect females quite frequently and often touch females on the hip to get them to stand in the right position for mounting. The female presents by orienting her posterior towards the male, and frequently when he has mounted, she looks back over her shoulder and teeth chatters or grimaces at him. In many species the female also gives a 'copulation call' as the male is mating, which may indicate orgasm on her part (Burton, 1971; Allen and Lemmon, 1981). A fertile mount is usually indicated by the male's ejaculatory pause and the presence of white ejaculate on the female. Sequences of mounts may continue over a period of several days until one partner or the other terminates the consortship by moving away from the other.

In some cases females do not cooperate with mating attempts by males and will not stand for mounting or remain in spatial proximity to them. In other cases males will not mount a female even though she may present repeatedly and try to drag him to his feet. Positive consort bonds seem to occur between animals who have some previous amicable experience of each other as co-troop members. Newcomers to groups must usually establish positive affiliation with a female before she will breed with him. Under certain conditions, common langur females may mate with a new troop male in order to establish a positive relationship with him, even if she is already pregnant (Hrdy, 1977).

The establishment of associations between non-related animals in a non-breeding situation is also an important social relationship in some species, such as rhesus macaques (*Macaca mulatta*) (Reynolds, 1970) . It is a particularly valuable source of support for incoming animals who can establish bonds with a resident. Immigrant male baboons (Smuts, 1985) who can establish positive social relations with resident females, frequently sit with them and hold or carry their offspring, thus providing protection and an extra source of socialization for the infant, as well as a positive social force for themselves (Ransom and Rowell, 1972). Spatial association, frequent grooming, friendly approach, relaxed sitting and support in agonistic encounters are all indicators of such friendly relationships which are evident to other group members. These bonds are so predictable that challengers will sometimes harass a male's female friend in order to force him to interact if he does not respond to direct challenges (Smuts, 1985). Another important aspect of interaction is the process of reconciliation after a fight or agonistic encounter. Unless the purpose of the fight is to drive the loser from the group, the two must reconcile to some extent in order to maintain group stability (de Waal, 1989). This may involve the loser presenting to the winner and being mounted, closely followed by an intense grooming session. Observations of macaques and chimpanzees suggest that a third party may be involved to bring the two combatants together (de Waal, 1982, 1989; Burton, pers. comm.). Sometimes if one animal has been chastised in a control interaction, the control animal will then mount it apparently in an effort to restore good relations between them.

Immediate Messages

All of these types of relations - socializing, kin, mate, friendly, reconciling - utilize a wide range of communicative postures, gestures and vocalizations in establishing and exhibiting special dyadic relationships. The same types of gestures are also used in the day-to-day transmission of information and attitudes which forms the basis of group living. Through the four channels of communication, individuals are informed of external facts such as the presence of food or danger, of internal facts such as estrus and state of health, and of social desires such as friendly interaction or reassurance. The major function of these messages concern social integration, and to a large extent are directed towards established social partners of one category or another. However, animals must still move through the group and respond to immediate interactions of approach, intrusion, displacement and resource acquisition. Since spatial position is such an important indicator of relationship, it is vital to understand the tolerance of other individuals to an approach, and to keep in mind that this tolerance may vary due to daily stress. Motivations to approach other animals also vary. A female with a new infant is often approached by both males and females who would not otherwise associate with her. She may also move into the 'nursery group'; a spatial association of mothers with young infants, and perhaps some interested juveniles or males. If food is becoming scarce it may be more difficult to eat in peace without other individuals displacing an animal from a food source it has found or prepared. Mothers must also be very watchful over their offspring and rescue or support them if they get into agonistic encounters. In some cases, mothers have been observed pulling their youngsters away from situations in which they might antagonize another adult and then come running to mother with an irate adult pursuing them (de Waal, 1982). Infants, juveniles and adolescents gradually develop their social bonds through play and exploration of their environment. Play fighting is distinguished from the real thing by the use of the play face and the relative silence of the activity. As soon as one participant gives a squeal of pain or alarm, an adult intervenes to stop the bout. Thus young animals learn how to indicate that what looks like a fierce attack is in fact play, and how to suppress vocalizations. It is clear that, broadly speaking, messages of group wide relevance, the establishment of long term social relationships and the transmission of immediately relevant information have a major effect on how a primate group operates. In order for group wide messages to be useful, the group must maintain sufficient spatial cohesion to receive the messages. The actual spacing will depend on group size, habitat factors, and the channel through which the message is sent. Maintenance of group boundaries is important since primate do not have positive relations with every conspecific that they see.

The structure of individual communication patterns is affected by the level of influence each individual commands in the group. This is demonstrated by the differential effectiveness of animals filling the various roles necessary for troop well being. The establishment and maintenance of these social relations is a life-long process for every primate. Mother-offspring bonds, kin relations, breeding associations and friendships are established, expressed and cultivated by communicative behaviour in all channels through the process of individual recognition, affiliation and mutual support. These very general germs come down to day-to-day interactions of grooming, sitting with, approaching and avoiding others, all with the appropriate visual, olfactory, tactile and vocal cues. The process of learning to live within and utilize the group system is the major task of socialization for an infant and to a lesser extent for an incoming migrant.

Coding

The coding process is very complex because it integrates the channel of the message, which includes both basic and interactive information, factors affecting the ability to code, the process of decoding which young animals and migrants must learn, and the influence of the maturation process both on the basic and interactional levels of message transmission. Thus the patterning or coding of the various systems which express information is a complex learning task, in order to integrate internal state, method of getting the message across, and the social consequences of so doing, into one coherent functional activity.

Channel - olfactory

The communicative aspects of scent-marking behaviour involve both the basic level of actual olfactory cues, and the interactive aspect of where and when the scent mark is applied, and under what circumstances. In addition to the olfactory cue is the visual information provided by the bodily positions and movements which indicate that marking is taking place. Most studies present objects which have been scent marked to test animals, and their responses are noted. Differential responses to various marks are considered to indicate that the test animal is capable of discriminating other individuals from their scent alone (Epple et al., 1982).

Usually there is a time delay between the act of marking and its discovery, which means that the ability to decode the mark in the absence of the sender is a very important skill which must be learned in the social context. The major choices here are those of location and frequency of scent marking by individuals. Scent marking has a variety of functions in delimiting ranges, scenting arboreal trails and allowing solitary animals to communicate their sexual state. Lemurs *(Lemur* spp.), sometimes vie with each other by over marking each other's marks, and by engaging in 'stink fights' in the presence of receptive females (Jolly, 1985). The angwantibo *(Arctocebus calabarensis*), has a scent which is described as very unpleasant and which functions to indicate to an infant angwantibo that it must remain immobile (Manley, 1974). Gorillas also have a fear scent which remains in a locale even after the gorillas have departed (Fossey, 1983). Even though these last two uses of olfactory communication are not voluntary, they are cases in which a particular type of social information is transmitted through the olfactory channel, to animals who are familiar with the meaning of such an odor. Thus, although the actual scent of an animal is largely a 'basic' level of communication in the sense that the animal has no control over what it smells like, various social and emotional messages can be gleaned by a receiver with sufficient social experience of the sender.

Channel - Auditory

Primates use sound for long-distance information transfer, as well as to express themselves to those nearby. Even among humans some vocal sounds are involuntary exclamations, such as cries of pain. Non-human primates also exhibit some of these uncontrolled vocalizations, such as infant shrieks and screams of terror or pain. There are also more controlled calls which provide information about external objects, such as food barks or predator warning calls of various kinds (Struhsaker, 1967). Grunts, lost calls, high-pitched vocalizations and location calls are all given under particular circumstances and indicate information about various individuals. Recent experimental work with play-back recording indicates that in some species vocalizations can be attributed to specific individuals. Work with vervets *(Cercopithecus aethiops)* (Cheney and Seyfarth, 1982), macaques *(Macaca fuscata)* (Green, 1975; Gouzoules et al., 1984), mangabeys *(Cercocebus albigena)* (Waser, 1977; Seyfarth, 1987) and chimpanzees *(Pan troglodytes)* (Marler and Hobbet, 1975) demonstrates specific responses indicating individual recognition. Mangabey troops approach or play back the call of their own group male, but avoid the call of a neighbouring one (Waser, 1982). Frequency and acoustic pattern analysis of certain types of calls indicate that sounds which are very similar to the human ear have differences which are correlated with different patterns of use. Japanese macaques *(Macaca fuscata)* (Green, 1975) discriminate between the 'coo' calls used by infants apart from their mothers and those used by sexually receptive females, even though the difference is not obvious to human listeners. Other species also make acoustic discrimination within particular categories of vocalizations that had not been discriminated by humans. Vervet grunts also indicate information about at least four types of external situations (Seyfarth and Cheney, 1986). The animals refine their system of vocal expression in terms of initial frequency, length, patterning and shape of the call, as well as the level of noisy overtones present.

Auditory recognition of individual difference is demonstrated in studies of various species such as chimpanzees, vervets and rhesus macaques *(Macaca mulatta)*. Bruno, who is one of the signing chimpanzees, was able to indicate whether taped calls were made by the same or different individuals, even when he was not familiar with the animals (Fouts, 1974). Vervet females have indicated that they can not only recognize the screams of their own offspring, but can also associate the calls of other juveniles with their mothers (Cheney and Seyfarth, 1982). Obviously this type of recognition takes time and experience to perfect. Some initial work on social relevance of auditory cue factors has been undertaken in rhesus macaques by Gouzoules et al. (1984). This indicates that the rank of the receiver and the contact/non-contact parameter influence the structure of the calls. Individual mothers who heard such calls given by their adolescent young responded to them differentially. Future research in which the relevant parameters of calls are separated and tested for their social and functional associations may reveal that a wide variety of choices are possible in the construction of vocal call.

Channel - Visual

Aspects of communication codified by gesture, movement, and colour utilize a range of components in the face and body. In the species *Macaca sylvanus*, I have examined the range of components used in facial gestures indicating threat, friendly approach and fear. The actual components used seem to be very similar in all three categories of expression although the combination and frequency of use may differ. Some indications of expression, such as eyes staring, eyebrows raised and ears forward, are common for all categories whereas eyelids flicker is more common in threat than in the other two expressions. Although there were 33 components of the face and head used by most members of the group, each individual had a different pattern of which ones it always used and which ones it used in some cases and not others (variable use). There was only one component used by a single individual, and this was not frequent. Thus it seems probable that senders utilized choice in composing the facial gestures both in terms of components used, their frequency and how they were combined. This variability underlies different types of messages and may mark differences between the individuals sending them.

Learning

The wide variety of functional coding in all three channels discussed will not affect the functional repertoire of communication unless the young become proficient in learning the code systems. Infants are born with the ability to make a few of the species-relevant vocalizations and facial gestures, but unless they learn how and when to use them they are not able to communicate effectively. Facial gestures appear to be even more complex than vocalizations initially, since more parts of the body must be controlled before they can be given accurately. Infants do get some leeway in the learning process and efforts to make appropriate gestures, such as the lip smack, are responded to by adults.

> Each time the infant 1 makes a sucking motion, both parents immediately chatter to it. By the third day at the earliest and the fifth generally, the infant 1 is capable of returning weak chatter movements when it is chattered to, but invariably just moves closer to the animal chattering to it. That is, it returns sucking motions with an added component of pulling back the lips and opening and closing the mouth. Thus a preexisting action movement, already included in the repertoire by birth, is conditioned into a purely social gesture. (Burton 1972: 33)

Infants who are raised in social isolation and do not have learning models to follow or social reinforcement, do not learn to code or decode the communication system (Sackett, 1968). Young who are raised together in peer groups do somewhat better, but they are still constrained in their ability to integrate into a social group (Suomi and Harlow, 1972). Experiments in which young animals are allowed to preferentially approach their mothers or other females indicate that young monkeys can recognize their mothers at a very early age (Nakamichi and Yoshida, 1986). In baboon groups where the mother has particular male associates, the infant also learns to discriminate this male from others and to respond to him with approach rather than avoidance (Ransom and Rowell, 1972; Smuts, 1985). Some infants have their own male friends who associate particularly with them and not necessarily with their mothers. They also show differential response to these animals as well (Smuts, 1985). Thus infants are early able to make individual distinctions and begin to learn the constraints of their communicative system. In some cases where infants are developmentally handicapped, they do not learn the system at age-normal rates, and may in fact be incapable of forming particular sounds or facial expressions. In situations such as this, especially if the mother continues to protect the youngster, most others tend to not respond when inappropriate gestures or responses are made. However, as an extension of this, other animals may not interact much with the handicapped one who may exhibit great distress when its mother moves out of sight (Fedigan and Fedigan, 1977). A four-year-old *M. fascicularis* I observed, shook with some type of palsy and was treated like a young animal since he was not punished like other four year olds for failure to respond to adult messages. For example he did not move out of the way of an adult male approaching along a narrow branch. After facially threatening the afflicted adolescent several times, the adult jumped over him and continued on his way, rather than pushing or biting at him. All other adolescents in the group quickly moved out of an adult male's path under similar circumstances without waiting to be threatened.

In addition to age and social maturity, factors of rank and kin relations are suggested as underlying communicative effectiveness by influencing the ability of the sender to code the message and the receiver to decode it. According to Miller's work (1967, 1972, 1975), the type of relationship between sender and receiver is an important aspect determining how well the message gets through. Unrelated conspecifics faced with a two-choice task did little better than chance in responding appropriately to a facial gesture. Mother-and-son pairs, on the other hand, were correct in about 90% of their responses. He also found that high-ranking animals were more efficient senders and less efficient receivers while lower ranking ones were the reverse. This may result from the direction in which attention is usually focused between disparately ranked animals.

Decoding

Thus the learning of the coding process is an ongoing process based on many individual factors such as age, life experience, rank, kin group, and individual abilities. Therefore some individuals will be more adept at sending precisely the message they wish to send than are others. To some extent the skill in coding must be matched by an equivalent ability to decode messages in order to comprehend (and act on) the layers of meaning packed into them. Decoding skills are based on recognition of individuals which may be influenced by group size, learning the social relationships between them, and on understanding the actual message being sent. Primates occur singly, in pairs, family groups, and troops ranging up to several hundred in multi-male, multi-female associations of, for example, Japanese macaques. For most species the upper end of group size range seems to be about 60 animals and most species have a smaller average group size. This range of up to 60 animals seems to be well within the capacities of group members to recognize individuals. Even the larger groups do not appear to malfunction due to difficulties in learning individual animals, but often large groups subdivide into clusters such as matrilineages or breeding subsets which may make the recognition process easier to learn. Learning of identities must be undertaken by the young and by migrants into the group. The basic cues provided through olfactory, auditory and visual channels are undoubtedly the foundation for individual recognition through age, sex, size, physical characteristics, smell, sound, and other individuating features. Infants return to their own mothers from a play session even at a very early age. Young animals also learn to recognize particular males with whom they and their mothers have positive social relations and will interact with them rather than showing the more common pattern of male avoidance (Altmann, 1980; Smuts, 1985).

The coding evident in communication transmissions of *M. sylvanus,* the Barbary macaque, enables individuals to decode the social identity of the sender (as opposed to the individual identity) as well as receiving the information of the message. The system appears to operate by differences in the occurrence and frequency of components such as eyebrows raised or ears forward in threatening and friendly gestures (Zeller, 1986). This study (Zeller, 1986; 1987) revealed that patterns of component use distinguished categories of age, sex and kin group. Older individuals tended to use more components than did younger ones, and also to use particular components at a higher frequency. Males and females could also be distinguished by the frequency of particular components in their facial communication patterns. Even more useful for understanding the social network is the discovery that maternal kin groups utilize facial components in a patterned way allowing a high probability of distinguishing between kin groups on the basis of the form of facial gestures in threat and friendly approach situations. The kin group-data is difficult to corroborate with other studies, due to paucity of data on kin grouping, but component patterning by age and sex is also indicated in research on Japanese macaques (*Macaca fuscata*) (Zeller, 1987). Therefore both young animals growing up in the group, and incoming migrants, have the opportunity to assess this information on social categorization and to respond to it appropriately. The kin-group indicators would be particularly useful since that data is much less obvious from physical appearance.

The nature of the components used to send a message may also indicate something about the rank level of the animals since the highest ranking ones tend to use rather abbreviated gestures, whereas the lower ranking ones often have to continue the gesture for longer, or utilize more components or those of higher intensity to get the same level of result (Zeller, ms.). This statement reflects a tendency rather than a clearly observed significant difference of the type found for class differentiation. The type of response engendered by a message will also affect its level of intensity and the number of components used to send it. A prompt response will generally cut off the transmission of a gesture, whereas being ignored, or responded to inappropriately, will often bring an escalation of the gesture. High-ranking animals frequently send low-intensity gestures, which can escalate from a brief threat to a full-scale chase if they are ignored (Reynolds, 1970). If both animals threaten each other, a bluff encounter or a fight can ensue.

Intensity of gestures can also be influenced by the social nature, as opposed to the response level, of the recipient. Infants sometimes ignore threats which then can become quite intense, but there seems to be little retaliation even if the threats continue to be ignored. As mentioned above, some handicapped young individuals are not treated like their age mates but like younger animals whose lack of appropriate response can be tolerated. Intense friendly lip smacks are also directed towards very young infants, and if they show signs of returning the gesture, the intensity can become very high and show tongue protrusion and hand clasping, as well as jaw and head movements in macaques (Burton, 1972). Occurrence, frequency and patterning of component use in gestures indicate individuality, social relations, rank, and some information about the level of response expected. A migrant entering the group can observe component use, gesture direction, intensity, speed of response, and type of retaliation in addition to watching grooming interactions and support affiliations. These types of data will allow the stranger to assess relationships between animals and thus what the best approach to integrating into the group might be. Migrants must become familiar with the social network very quickly in order to avoid making social errors in their efforts to form affiliative attachments. Also, they need to learn about the relationships among animals in order to recognize which animals must be responded to in the presence of which other ones (triadic or supportive interactions). The complex web of kin support, friendship and affiliative solicitation that surrounds each animal, makes communicating with it a process requiring judgement and knowledge of the other's support network. A migrant who cannot pick up the type of information very quickly is at risk of having his attempts to integrate himself into the group rebuffed more and more strongly, until he must leave. It is quite possible that this is why migrants sometimes stay on the fringe of a new group for some time before attempting to interact with central animals. Also, moving into a group with at least one familiar animal can potentially provide some support if other animals respond negatively to the migrant's approach. The social need for some friendly interaction which may translate to social support is probably also a reason why incoming animals try to affiliate with one particular individual (such as a female with infant) early in their integration process.

Successful decoding depends on the abilities outlined above of recognizing individuals, perceiving social relations, and understanding ranking systems. Just as young deprivation-reared animals could not code their communications successfully they usually cannot decode the messages of others when placed in a social situation. Sackett's (1966) studies of isolate-reared infant macaques suggested that they responded appropriately to yawn threats, although these tests were not carried out under social conditions. But simple recognitions are not enough to allow an isolate infant to interact successfully in a group. They often exhibit strange behaviours and respond inappropriately to communications of others. Even peer-reared animals are not very successful at complex communication when they are introduced to social groups. These animals can make barks and high-pitched vocalizations as well as threats and grimaces, but they have very little idea of how to apply them socially, or respond to others sending gestures. Adult failures to decode messages may be involved in the deaths of young orphaned-group raised infants who are adopted by males who eventually attack them because they shriek from hunger (Taylor et al., 1978). Conversely, "baby-sitting" of an infant by an adult of another group or species permits transmission of specific or local codes. Learning how to decode messages is thought to facilitate hybridization in polyspecific groups (Burton and Chan, 1987).

Another way in which decoding can be examined is to look at patterns of deception and bluffing. Deception is basically communicating something that is not true (Whiten and Byrne, 1988), and acting in such a way as to bring about a non-obvious result. The success of a deception is directly related to a primate's ability to send a message at variance with reality and have it accepted. The study of coding is greatly facilitated by examining situations in which it does not work and the deception is discovered. Both the production and the recognition of deception imply a mastery of the coding and decoding process. In fact, the manipulation of an adult defender by a juvenile who gave an unwarranted distress call which provoked the defender to attack a nearby adult female is one episode of deceptive communication which I have witnessed. This was repeated several times but when the defender was well out of sight the juvenile was grabbed and shaken by the implicated adult female. Bluff is used under certain agonistic social situations as when attempting to intimidate a rival or enter a new group without having to fight. The decoding question in a bluff situation is how well does the bluff - the indication of confidence and ability - actually correspond to real ability to do damage in a fight. It is important for animals to be able to differentiate between the message and the real situation so that they do not either back down every time and risk losing opportunities or get embroiled in too many fights in which they are outclassed.

It is clear from the preceding that decoding occurs at several levels, particularly those of message structure, memory and learning. Animals must learn the meaning of gestures, individual differences in message styles, the patterns of social coding, the influence of rank on the system and how to distinguish false messages from true ones. It is the reciprocal process to coding, but the two processes are separate and are not necessarily accomplished with equal ability as is demonstrated in Miller's work (1967, 1972, 1975; also Woodruff and Premack 1979).

Maturation

Up to this point I have mentioned the aspects of communication in which learning and a socially effective upbringing influence the ability of a young animal to master the system. There are also aspects of physical maturation which are quite relevant to the process of transmitting information. These mainly involve the message properties of various physical systems which develop over the course of an animal's life. It may be argued that these are more on the level of individual markers, but some features of size and sexual characteristics have social relevance as well. For example, the transition from one age category to the next is partly due to physical maturation and partly due to social acceptance of the change. Young male baboons can be fertile in their subadult phase, but seldom take their place in the breeding and status interactions until they have grown their manes and approximately reached their adult size. Two *M. sylvanus* males who became leaders and breeding males at a fairly young age did not grow as tall as their age mates, but increased in bulk, shoulder size and testes size, while cohort males gained more height, but not the bulk and secondary sexual characteristics of these socially more mature animals (Burton, pers. commun.; pers. observ.). Throat sac development in calling primates or degree of hyoid bone development in howler monkeys *(Alouatta sp.)* will influence the types of calls the individuals can make and thus reveal both age and social status information. Sexual behaviour is important as a correlate of age, since subadult animals frequently make the transition to adult status when they bear or sire offspring. Engaging in sexual activity requires both the ability to recognize the other sex and the knowledge of the appropriate communicative behaviour to elicit responses in the other animal. Males must learn to solicit a female's favour and this may mean competing in displays with other males (e.g., stinkfights) as well as indicators of interest such as 'bird-dogging' seen in Japanese macaques *(M. fuscata)* (Fedigan, 1982). Females of many species also utilize proceptive behaviours to interest males in sexual activity such as presenting, staring, vocalizations and snag crashing as seen among orangutans, as well as direct genital stimulation.

Physical information is usually easy to perceive - size, colouration, sound, smell - but requires knowledge of the other as an individual and social being in order to fit the cues into a useful framework of what smell means which animal. This information in which physical development has both individual and social meaning is part of the reason that the social interactions of strangers are so abbreviated and tense. If positive social interactions are established the animals will often spend a lot of time in proximity, grooming or sitting, and will rapidly learn to predict the other's responses to various stimuli. The system works so smoothly in most long-term social groups that it is often difficult to realize the layers of information displayed and comprehended in the simplest social interaction. Having undergone the process of physical and social maturation animals exist in relation with potential mates, peers, offspring, rank systems, friends and enemies, and have patterns for communicating with all of them.

Conclusion

The transmission of social data via the coding process is a level of information transfer which has only recently been recognized as being a component of the message being sent among non-human primates. It is rather like status markers in human speech, in which various dialects and word uses indicate different social standings even when the substance of the message is the same. The complexity of the system is one reason why even closely related groups have differing patterns of component use. For example, Japanese and Barbary macaques who are at the geographical extremes of macaque range, use many similar components in their gestures, but combine them in different frequencies and patterns (Zeller, 1987). Until fairly recently many people considered that the form of macaque gestures was fairly standardized, and this is true to the extent that many of the gestures can be recognized, but the social overlay which is reflected in patterning of component use is not the same. Appreciation of the pragmatic and social functions of communication reveals that learning the whole system of codes and channels and being able to function in a unified communication system is a feat requiring a considerable amount of learning. I have seen a young macaque off, by herself, making various facial gestures, going through the movements in a non-interactive setting. Learning is reinforced strongly as in the response of mothers and other adults to an infant's first attempt at a lipsmack. Animals crowd around, crouched low, lipsmacking with great energy to the infant, often using a headshake and hand clasp which indicate the intensity of their positive reinforcement of these first social signs. This functional approach to communication systems gives a much different view of how the process works than do studies of channel and form. Nonetheless, channel and form are important aspects to take into consideration in addition to complex social factors, when attempting to understand what communication is, and how it works in the social units of non-human primates.

CHAPTER FOUR

Baboon Social Communication: The Social Contingency Model and Social Exchange

by

J. K. Chadwick-Jones

The influence of social communication on the development of cognitive capacities of nonhuman primates has been highlighted in several classic discussions (Chance and Mead, 1953; Humphrey, 1976; Kummer, 1978, 1982). While the insights and conclusions of experienced researchers suggest that capacities actually used in social interactions are complex (Rowell, 1972; Savage-Rumbaugh, Scanlon and Rumbaugh, 1980; Burton, 1984; Walters and Seyfarth, 1986), their systematic treatment is at a tentative stage, with many controversial questions to be addressed (Byrne and Whiten, 1988a, 1988b).

One current approach to the understanding of nonhuman primates is to analyze their behaviour exactly as one would analyze human social skills, using theories, models, and techniques that are used in social psychology (Chadwick-Jones, 1987a, 1987b). Techniques and ideas that may improve the empirical grounding for theory by suggesting new areas for research have occasionally been borrowed in this way, for example, by Sade (1965, 1972) and Reynolds (1981). In social psychology, actions between one individual and another are studied over the short term and the aim of proximate theory in this discipline is to clarify these immediate effects. To assuage any disquiets that zoologists might have about this frame of reference, let us note that identifying proximate social process can strengthen the functional theory (Dunbar, 1988). So, in any case, theories at the proximate level and at the ultimate level (theories of reproductive fitness) are complementary (Jamieson, 1986).

Social skills analysis (Argyle, 1988) in humans requires close attention to elements of detail, especially from non-verbal communication, and this seems an ideal frame of reference for studying the rich social contacts between baboons that have been extensively described by Kummer, 1968, 1975; Ransom, 1981; Abegglen, 1984; Dunbar, 1984; Smuts, 1985; Strum, 1987 and Colmenares, 1990. A substantial amount of information exists which will permit us to focus on detail that is needed for a social skills analysis. In other monkey species there are also rich empirical sources, such as those provided by Bertrand (1969) for stumptail macaques *(Macaca arctoides)*; by Struhsaker (1975) for colobus monkeys *(Colobus badius)*; Rowell (1972) for rhesus monkeys *(Macaca mulatta)* and by Zeller (1980) for Barbary macaques *(Macaca sylvanus)*. The results of many of these studies are well-summarized by Fedigan (1982).

In several of the issues touching on social behaviour that are discussed below, mention must be made of the immensely important studies of chimpanzees *(Pan troglodytes)* by de Waal (1982) and Goodall (1986). There is no sharp dividing line, as social animals, between humans and nonhuman primates. Even if the evidence for a closer evolutionary relationship is much stronger between humans and apes (Lowenstein and Zihlman, 1988) the capacities of monkey species are worthy of attention because of their greater numbers and availability for study and especially because they establish the prior argument, that is, a baseline of capabilities before the emergence of hominoids (Burton, 1984).

Communicating Intentions

The first topic for discussion here will be the use of signals in baboon communication and an analysis by a social skills approach borrowed from social psychology. The second topic will be the possible applications of social exchange theory to baboon social behaviour and I will try to show the value that these applications may have for extending our knowledge of baboons' social intelligence.

Baboons are extremely versatile in their use of signals to communicate their intentions and emotions to others. Some idea of this is to be gained from Hausfater and Takacs' (1987) study of signals involved in the presenting posture among yellow baboons *(Papio cynocephalus).* Presenting (turning the rear towards another individual) gives an excellent illustration of how variations in combined signals can convey a message. Hausfater and Takacs showed how body posture, tail position facial expression and vocalization conveyed different messages in several categories of presenting that will be discussed below. Presenting is not at all a stereotyped body posture, it is a flexible set of variations according to context and intention. For example, presenting-for-grooming may vary across present-head, present-shoulder, present-side, as has been recorded for geladas (Theropithecus gelada) by Dunbar (unpublished data).

In its origins and in its major usage, presenting is a sexual posture in the female prior to the copulatory mount of the male but it appears in a wide variety of asexual contexts (Chadwick-Jones, 1989) as a signal between males, between females or by a male to a female. For example, presenting may be used as a request for grooming or a request for permission to approach another, to share a location or even food with that individual. Between males, presenting is a step towards forming a friendly relationship and in this case is followed by grooming. Presenting, therefore, is a tentative signal: an indication of friendly intention or an invitation to groom and some investigators see it as an important signal in a negotiation process, typically between male baboons (Colmenares, 1990).

Inferring negotiation between male baboons implies that individuals have "plans" at least over the short term. Kummer (1982) has commented on the insights of field observers that agree with the notion of individual "plans" but he has mentioned, also, the difficulties of finding empirical support for it. Sometimes, episodes where intentions seem indisputably present are reported as anecdotes that do not have sound empirical status because they *are* asides, noted in passing, while the focus of systematic study has been elsewhere, on foraging strategies, ecological or reproductive issues.

Social Contingencies

Recently I have examined several reports of field observers whose work treats the social behaviour of baboons as a central issue - namely, the studies by Ransom (1981) and Smuts (1985). From their detailed field reports can be confirmed that there is continuity in the behaviour undertaken by an individual in relation to another over a period of time. I have transferred some of this observational data into the format of the social contingency model to provide a simple, continuous record of sequences in the actions of one individual and the responding or initiating actions of another (Chadwick-Jones, 1991). This model is taken from social psychology where it was first proposed by Jones and Gerard (1967).

The basic idea is to record every signal (action) by each of the pair in a sequence that depends partly on each partner's own previous action and partly on the action of the other. The model looks like a zig-zag diagram of a familiar kind that Tinbergen (1951) used for stickleback courtship patterns, but there are important differences. Tinbergen's diagrams were idealized in the sense that they did not include either rare actions or repetitions. In contrast to this, the social contingency model is intended to accommodate actual step-by-step sequence; it represents the empirical data of actual sequences. Another major difference is that the social contingency model suggests a typology - social episodes can be classified insofar as they involve separate or joint objectives of the pair. In its applications in social psychology, the model has been used for training in social skills by encouraging attention to the most basic elements of face-to-face communication; it has been used in therapy to introduce an increased awareness of the effects of gestures and posture (Argyle, 1978, 1988).

The classifications are as follows: the reactive contingency (for superficial brief episodes with no long-term plans); the mutual contingency (for the development of joint plans achieving joint objectives); the asymmetrical contingency (one of the pair follows a consistent strategy and has an objective, the other copes with it). In humans a reactive contingency might be a brief, casual conversation; in a mutual contingency both partners have plans and their actions match and promote each other's plan. The asymmetrical contingency might refer to a dominant individual's imposition of goals on another. Inferring a plan becomes possible from observing consistency in a sequence of recorded actions. The point of view taken here is similar to the anthropologist's "etic" position, applying a conceptual framework that is meaningful to an observer (Harris, 1987).

The Conflictive Contingency

The model has been augmented, for baboons, by a fourth category, the conflictive contingency, where the objectives of the partner clash and the conflict will resolve eventually in favour of one of the two. The conflictive contingency comprises negative sequences where the partner's objectives are in opposition to each other. The signals are mutually threatening, as in a contest over dominance; the contest can take place through mutual posturing and threats. The strategy of a male who is challenging another's dominance can be understood as 'putting down' the other individual and seems analogous to human strategies, although this kind of negative sequence has rarely been studied in social psychology.

'Winner' and 'resister' strategies emerge from Ransom's (1981) account of a dominance contest between a male leader and another male who enters the periphery of the group and, over a period of months, makes a prolonged challenge, continually following (and menacing) the leader's dominance, finally displacing him. This sustained challenge, a kind of 'wearing down' strategy, suggests a capacity for holding to a consistent plan.

Fig. 1. Conflictive Contingency

Figure 1. CONFLICTIVE CONTINGENCY

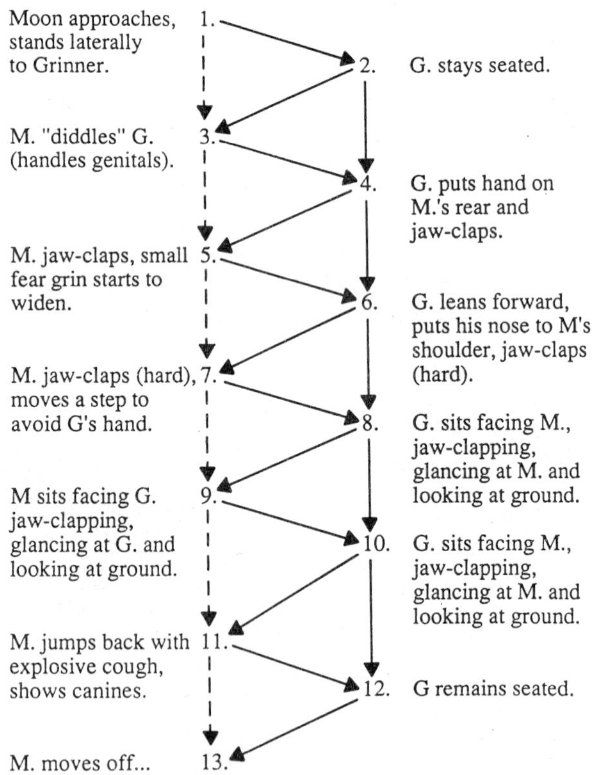

Ransom described how the challenging male (Grinner) harassed the leader (Moon) in a subtle fashion. Grinner's location and presence seemed to carry most of the message, there were no overt threats other than jaw-clapping. In contrast to this, Moon was excited and nervous, showing this in a fear grin, explosive cough and by jumping back. In the episode shown in Fig. 1 there is also mixing of signals: both animals have touched each other, usually a reassuring gesture, limiting the possibility of physical aggression, and opening up the possibility of affiliative relationships.

The Mutual Contingency

Another brief extract from Ransom's observations of olive baboons is given in Fig. 2 (Mutual Contingency).

Fig. 2 Mutual Contingency

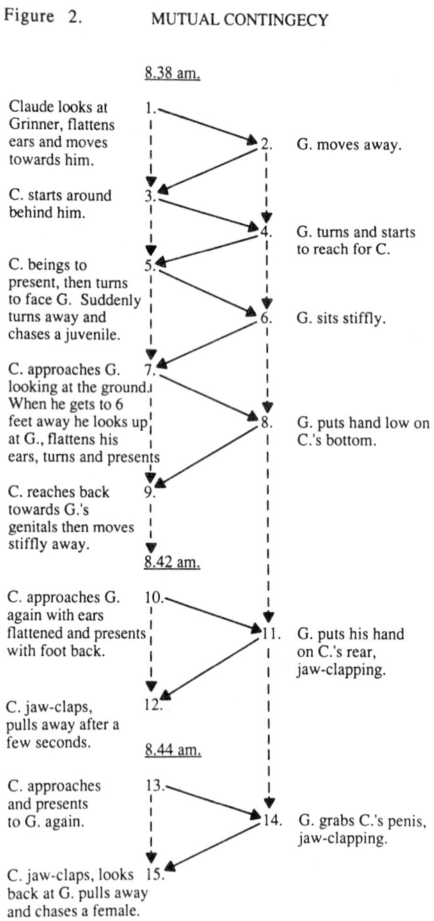

This is the relationship between Grinner and a younger male (Claude) who makes a series of friendly approaches to which Grinner responds similarly. The pair moves towards a settled dominant-subordinate relationship in a mutual contingency where joint objectives are achieved, insofar as Claude signals acceptance of the dominant status of Grinner and both display friendly gestures. Ear-flattening is a signal that is repeatedly used and seems to express Claude's fearful approach to Grinner. Hall and Devore (1965) in their studies of chacma baboons *(Papio ursinus)* and olive baboons *(Papio anubis)* classified ear-flattening as a friendly gesture. Colmenares (1990) in his study of hybrid hamadryas and yellow baboons *(Papio hamadryas* and *Papio cynocephalus)* at the Madrid Zoo referred to ear-flattening as a reassurance gesture by individuals who were either greeting, soliciting or giving support or who were unwilling to attack. In Fig. 2, Claude seems to be negotiating a settled relationship with Grinner by tentative presenting. In these sequences, the action of presenting is a decisive indicator of Claude's friendly intentions. Both Hall and Devore and Colmenares mentioned that ear-flattening could also be a threat-gesture and more information is needed on variations in the message. Future use of the model could provide this information along with information on other signals such as eye-contact.

The complex uses of signals is well-illustrated in Hausfater and Takacs' (1987) study of presenting in yellow baboons. They proposed six classifications according to the message: submissive, sexual, mixed sexual-submissive, affiliative, abbreviated (characteristically a permission-seeking gesture when approaching close to a dominant) and presenting-for-grooming. There are few descriptions available of the range of presenting postures and of the signals that accompany presenting. Hausfater and Takacs mentioned that submissive presenting involves raising hind-quarters, grimacing, high-pitch vocalizations, cowering, tail twitching, tentative glances. Their focus on detail is similar to the social skills analysis of human non-verbal and paralinguistic signals.

Based on the descriptions at this level of detail, the social contingency model can be a useful source of information on baboons' intentions and plans because it has more empirical content than other forms of reporting. It tells us more about the use of signals in the forming and development of social relationships than is available from other sources and, potentially, can tell us more about baboons' cognitive capacities.

Analogies to Social Exchange

Ordinary language terms have inevitably been used to interpret the social actions of apes and monkeys, the most recent and systematic descriptions included. Ordinary language terms, also inevitably, carry the assumption that behaviour is purposeful, that individuals have intentions. This assumption produces what Asquith (1984) refers to as the "generic anthropomorphism" of language. In interpreting the social activities of apes and monkeys terms like "greet," "threaten," "deceive," are taken from ordinary discourse. There is no other way of explaining the proximate circumstances of primate social behaviour except by terms that are used for human social behaviour. It is appropriate that this should be so to the extent that what apes and monkeys do is often similar to human actions, but the work of explanation does not stop at this point. In attempts to achieve greater precisions, more specialized terms have been taken from the academic disciplines of sociology, social psychology, and social/cultural anthropology. Special terminology can be useful as a way to a better understanding than could be obtained through ordinary language, because definitions are agreed upon among practitioners of those disciplines on the basis of their observations of social behaviour within a stated theoretical framework. The transfer of terms from human social and life science is well illustrated in the case of social exchange theory (Homans, 1961). It has been especially useful in providing hypotheses about the significance of social grooming (allogrooming), a widespread behaviour in many species. The concept of an exchange of activities was first used by Carpenter (1942) who introduced a brief discussion of the "behavioural exchange" between mother and infant of mutually comforting behaviour. Peter Reynolds has argued that exchange provides a fundamental rationale for many primate activities and that relatively simple behavioural exchanges among apes and monkeys are forerunners of the complex networks of object exchange in human societies. Allogrooming, in particular, he identified as "an antecedent to social exchange involving objects," venturing the opinion that "human evolution has been characterized by the transformation of grooming networks into object exchange networks" (1981: 172). Several studies mention concepts familiar to social exchange theory and economics such as "payoffs," "balance" (Harcourt, 1978, 1979; Hinde, 1983) and cost/benefit ratios (de Waal, 1982).

Social Exchange Theory

Social exchange theory actually consists of a heterogeneous collection of explanatory propositions and hypotheses that assume reciprocal giving-and-receiving of actions. Statements about exchange are often controversial: for example, some authors emphasize economic models and profit-seeking while others interpret exchange in relation to rules of equity and generosity. Social exchange is defined rather differently in each of the disciplines of social psychology, sociology and social/cultural anthropology. The first of these focuses on individual motives (usually self-interest) for entering into exchange relationships. In the latter two there is more attention to the social system in which exchanges take place and to the social context as an influence on what is given and received, to whom and from whom (Chadwick-Jones, 1986).

The perspective on social exchange used in primatology (for example, in de Waal, 1982; Hinde, 1983) sees the exchange as a reciprocal transaction with the end result of gain, in some degree, for each of the parties (Mauss, 1925, 1954). Economic analogies are clearly suitable for this approach to social exchange and a terminology of rewards, costs, profits and opportunity costs has long been transferred from economics. If social exchange ideas are to be useful in primatology it should be because they identify certain kinds of information that otherwise would be overlooked by researchers and draw attention to cognitive capacities that otherwise would not be studied. For instance, a social exchange can be demonstrated only if it is established first that one animal recognizes another as a partner, with lapses of time between giving by one and a return favour from the other.

Obstacles to an Empirical Test of Social Exchange

Methodologically, it will be an exacting requirement to demonstrate the delivery of return favours over long time periods. There are two difficulties: first, there is the practical difficulty of observing two or more animals continuously over long periods of time. Second, how can the connection between an action and its return favour be established in conditions where many influences and other events are operative? How can it be shown that previous allogrooming and a later action are contingent items in an exchange, rather than behaviour emerging from kin relatedness, familiarity and locational proximity? Kin relationships could be controlled out by observing only unrelated animals but this severely limits the empirical sources; in many species, animals are almost always in proximity to kin and social bonds are formed with them.

Besides the methodological difficulties in monitoring social exchange, there are other, substantive, reasons for doubting that long-term exchanges can be demonstrated. Kummer has commented as follows: "So far, nothing suggests that grooming or any other social behaviour among primates is specifically timed as a post hoc reward for a specific action that was useful to the groomer" (1978: 46). Kummer does not believe that nonhuman primates are able to "integrate the objective value of a comparison across time and context to such a degree that the evaluation can outweigh emotional ties" (but see Burton, 1984). He acknowledges that the current lack of evidence might partly be the result of inadequate research methods which in essence could derive from a "biological tradition" of using the least onerous explanations and so overlooking what further evidence could be sought. The fact is that we have in the literature a large body of opinion suggesting persistently that social grooming entails exchange in some form and a few empirical sources confirming that grooming can entail an exchange, although these span only very short time periods.

To summarize, there are methodological obstacles for attempts to confirm a connection between grooming and other positive actions (such as helping, aid-in-combat, food-sharing) over the long term. There may also be doubts that nonhuman primates possess the cognitive capacities for rendering a behavioural *quid pro quo* at a later point in time because of some earlier action received.

Exchange and Altruism

When Packer (1977) wrote his classic account of olive baboons' reciprocating support in fights he did so in terms of reciprocal altruism and not as social exchange. One important difference between social exchange and altruism is that exchange does not involve risk, as altruism does. Social exchange, by definition, involves giving at relatively low cost for high subsequent returns; no risk of injury or hazard to the giver exists. As Trivers (1971) argued, reciprocal altruism (like exchange) should also incur low costs relative to ultimate benefits, but for *reciprocal altruism* these are to be assessed in terms of reproductive fitness. Explaining reciprocal behaviour as an exchange does not involve assuming consequences for reproductive fitness (although those consequences are not ruled out, they are not necessarily included in).

In social exchange theory, the focus of study is on the immediate context of an action and its reciprocal action. The length of time between these actions may vary - over days or even weeks and months. Although ultimate effects are not comprehended within the theory, nevertheless if the proximate causation is confirmed and is consistent with a functional outcome then the functional or ultimate (evolutionary) theory is also strengthened (Dunbar, 1984).

Cognitive Capacities for Exchange

As far as the cognitive implications of social exchange are concerned, several kinds of capacities may be suggested. The capacity to recall past actions and to deliver reciprocal actions is consistent with the circumstantial evidence of aid given to grooming partners (Dunbar, 1984; Smuts, 1985) but the causal connection is far from clear because of other factors, such as kin relatedness, proximity and familiarity, all of which could predetermine both allogrooming patterns and the giving of aid to others. The capacity to anticipate events of indefinite occurrence in the sense of "cultivating an ally" is indicated plausibly in observations of gorillas *(Gorilla gorilla berengei)* by

Harcourt (1979) and chimpanzees (*Pan troglodytes).* "Cultivating an ally" or commitment to a grooming partner may express cognitive strategies, but at the moment these are speculative issues. I have mentioned some of the methodological difficulties in confirming the exchange explanation and really we are left with a number of open and controversial questions, calling attention to new aspects for data-collection. The prize for future efforts might be validation of Reynold's (1981) view that behavioural exchanges in nonhuman primates are an antecedent of object-exchange in humans.

The Special Case of Negative Exchange

In social psychology, retaliatory actions have received very little mention in discussions of social exchange theory. Usually positive consequence are emphasized almost exclusively and the motivational roots of exchange have been found in reinforcement theory (Homans, 1961). A particular quality of negative exchanges, in agonistic behaviour, is the delay or time lapse between a hurt and its retaliation. Smuts (1985) observed retaliatory actions in olive baboons that were separated by time periods ranging from hours to days and she noted also that apparently unprovoked attacks were likely to have been retributions, spanning much longer time periods.

Allogrooming and Exchange

The explanatory language used by primatologists frequently suggests that grooming is a strong and direct influence on future reciprocal behaviour. For example, McFarland (1987) suggests that for the groomer the future benefit deriving from the grooming investment may yield future support as well as the basis for forming a coalition with the groomee; it is allogrooming that is the currency used to maintain the close bonds. Here McFarland is using an economic analogy of the kind often found in social exchange theory.

Among recent studies Smuts (1985) has probably made the most explicit declaration of a direct connection between grooming and future aid when she discusses the outcomes for baboons in male-female consortships. In these consortships females groom males much more than males groom them. Smuts (1985) asks why this should be and answers that perhaps females groom in exchange for some advantage from male proximity. She feels that the weight of evidence suggests that grooming may be employed to achieve social benefits.

Exchange and Alliances

In situations where grooming partners also help each other in alliances against third parties, social exchange is again a plausible explanation. But this is a largely circumstantial inference. True, also, that kin relatedness, proximity and familiarity may predetermine both allogrooming and alliances. But the social exchange hypothesis is persistently mentioned (Hinde, 1983). Dunbar's (1984) field studies of female geladas gave ample evidence of long-term support between grooming partners as did Smuts' (1985) study of olive baboons. The flaw in the connection, from the point of view of social exchange theory, is that interventions in fights with some risk of injury can be more suitably explained as reciprocal altruism. Social exchange could only be a preferred explanation if the risks were minimal: for example, if the opponent was weaker, the intervention infrequent and limited to threat displays rather than physical contact. Under these conditions and only if the grooming partners were unrelated should the social exchange notion be considered. Otherwise, for relatives, inclusive fitness would be the preferred explanation (Hamilton, 1964).

Relatives and Non-Relatives

Confirmation of a presumed exchange of grooming for aid-in-combat must be sought in the particular conditions where partners are not related, because coalitions with kin can be explained by the theory of inclusive fitness. In most species (except for chimpanzees, *Pan troglodytes)* non-coalitions are only a minority of the coalitions for which empirical data is available. Across guenons and colobines, macaques and baboons, the influence of kin relatedness is preponderant in interventions in fights. Individuals who intervene do so mainly on behalf of kin and are likely to undertake greater risks when they do so than when they intervene on behalf of non-relatives (Silk, 1982; Fairbanks, 1980). Harcourt and Stewart (1989) observed that the majority of interventions by gorillas were to protect kin and these were always against opponents dominant to themselves, whereas support for distant relatives was in no case given against a dominant opponent. De Waal's (1977) Long-tail or Java monkeys *(Macaca fascicularis)* supported non-relatives when they were threatening another but not when they were threatened by another, implying that they were more willing to join a non-related aggressor (at less risk) than they were to defend a non-related individual menaced by a possibly stronger opponent.

These observations highlight an important issue: coalitions vary in degree of risk and the low-risk coalitions should be our focus of study if we want to find support for the social exchange notion, although it should be noted that taking what Dawkins considers a slight risk can also be considered as altruistic (Dawkins, 1976). Silk (1982) has pointed out that coalitions between non-related female rhesus monkeys tend to be against lower-ranking females and therefore involve little risk of injury. These non-kin interventions do not fit so well with the notion of altruism (incurring risk) as they do with the notion of an exchange involving time and energy only. This fits with the basic "trick" of social exchange: that one partner should give, at relatively little cost, something that is highly valued by the other (Homans, 1961). This is a notion that is widely mentioned across disciplines when the exchange of mutual benefits is discussed. Under these conditions the grooming-coalition exchange is plausible. Grooming involves costs in monitoring and following the groomee (Dunbar, 1984) but these are likely to count little, if weighed against protection from harassment. On the other side of the exchange, grooming is a sought-after advantage; this much can be inferred from Struhsaker's (1967) reference to supplanting others from grooming relationships as a principal objective of dominant vervet monkeys. If interventions do not happen frequently, if they involve threat displays rather than physical contact and if opponents are weaker, then the groomee's benefits from continual grooming must be greater.

To summarize thus far, the essence of an exchange, the open secret of exchange (Homans, 1961), is to give what involves relatively little cost to the giver but is highly beneficial to the receiver. A coalition intervention that brings little cost to the giver but considerable advantage for the animal receiving aid, has the right conditions for exchange. On the other hand, if an intervention carries with it a high risk of injury, then the relationship will be better understood as altruism, or reciprocal altruism if the risky actions are reciprocated (Packer, 1977). Kaplan's (1978) study of interference in fights among free-ranging rhesus monkeys of known genealogy produced questions about support for non-relatives that Kaplan considered in terms of three possible explanations: Hamilton's (1964) theory of inclusive fitness was ruled out for non-relatives; Triver's (1971) reciprocal altruism notion was also ruled out for non-relatives because reciprocity of aid was low compared to reciprocity between relatives. Kaplan turned to a third possibility, that non-relative intervention could maintain affiliative bonds, thereby being "acts of affiliation" similar to grooming. But Kaplan had examined fight interventions only and did not include grooming or other forms of social activity in his data, and obviously, the possibility of an exchange between fight intervention and other activities was not tested.

Conclusions

I have summarized in this chapter some of the possibilities of research and a few of the questions arising out of the application of social-psychological approaches to nonhuman primates. In the social contingency model, a close look was taken at sequences of social actions. Here baboons' actions could be given as detailed a treatment as in studies of human social skills. In this way, we can learn more about baboons' intentions, about their "plans" in relation to other individuals, which may be sustained over weeks or months. More can be learned about baboons' capacities for thinking and following a plan as well as about their expression of emotions.

Secondly, an approach was made from theoretical ideas borrowed form social psychology. Here I used social exchange theory as a framework requiring careful attention to reciprocal behaviour and to the exchanges that can occur in a given relationship over periods of time. The notion of exchange is based on assumptions about cognitive capacities (as Williams, 1966, noted in the case of reciprocal altruism), involving recall over varying periods of time of previous transactions with particular individuals. Hypothetically, anticipation of future benefits is also a topic for our consideration. A sufficient number of open questions now arises to make both of these approaches from social psychology worth following.

Acknowledgements

I thank Les Barden, Medical Research Council Unit, Sub-Department of Animal Behaviour, University of Cambridge for producing the figures.

CHAPTER FIVE

Social Behavior in Bounded Space: Epistemological Issues in Producing The Texts and Textuality of Primate Behaviour

Susan Hornshaw

The theme of this volume is that the rules OF or FOR behaviour in groups of nonhuman primates are, essentially, "local phenomena with transcendent properties" (Burton, this volume). The rules governing a group may, in fact, be perceived as being group-constructed, as a function or derivative of interaction patterns among members. We have all been examining social behaviour in an attempt [1] to perceive the potential patterns of behaving in social groups of nonhuman primates, i.e., discerning the patterns OF behaviour and [2] to explain these perceptions of pattern. In the extraction of what we see as the nonrandom patterned statements about complex behaviour, or rules OF behaviour, we must, I think, be careful to remember that these "rules" are generalized notions about the nonrandomness of the behaviour of other species. These generalizations are most often statements that approximate the data of behaviour, i.e., they truncate the represented variability or variation of behaviour. These truncated statements as representations of the behaviour of primates ought not to be taken as stereotyped or reified notions that purport to describe the behaviour of social groups of primates.

I intend this chapter to be a discussion of how my research addresses the issue of local or "group traditions" of behaving in social groups of nonhuman primates. In addition, however, because information or knowledge is intended to be cumulative, this chapter also includes a reflexive analysis of the epistemological problems inherent in this research. As members of an intellectual and scientific discipline, we will make no progress in our understanding of nonhuman primates or our scientific explanations, if we do not subject our work to scientific scrutiny. There has always been a commitment to scientific scrutiny; what I am suggesting is that this reflexivity could be more broadly and intensively applied. I am not advocating a kind of public self-flagellation; what I am suggesting is that we really be our own best critics.

At the outset I must make it clear that most of my research has been with captive or managed populations of primates; consequently, my exploration of primate behaviour begins with a kind of theoretical sympathy for captive primates and those primatologists who work with captive primates.

The doing of science in a western cultural context means that we replicate and reinvent our cultural notions in general and our cultural notions of nature and natural processes in particular. One general view of nature and natural processes is that they are efficient or "economical." Our view of what constitutes a good or powerful scientific explanation is that it also be efficient and economical. There is a tendency toward explanations that have the appearance of being "simple" and parsimonious. We laud a kind of frugality in explanation that may have, as its ultimate effect, a denial of the richness of a primate behavioural data base. This frugality is encouraged when the core of an article becomes its statistically significant results. This is not an indictment of the primatological journals. For reasons of efficiency and costs of production and the volume of primatological research, they have encouraged authors to do this. By contrast, the classical essay format is not efficient for journal production, but, it does have its place.

We have become impressed by the form of the presentation of scientific laws. A striking example of this rather formulaic primatology is the work by J. Altmann on the "calculation of caloric expenditure in lactating baboon mothers" (Altmann, 1980: 53). This search for nomothetics has put us in a very dangerous epistemological situation. We really are looking for scientific "one-liners" that purport to explain the behaviour of nonhuman primates. We deploy the "one-liner" the way that we deploy the meaning of an arithmetic mean; i.e., ultimately we tend to ignore the range of variation in our search for the "single variable" that "accounts for most of the variance." We imply that this or that variable, like genealogy or gender as an example, is a statement of the nonhuman primate underlying code FOR behaviour. Often primatologists shift from using statements of description to using statements of proscription, i.e., the metaethical construction of nature or the construction of how nature "ought to behave" rather than a description of how nature, or, in this case, primates, do behave.

This procedure is exemplified by, for example, Ehardt-Seward and Bramblett, 1980; Rondinelli and Klein, 1976; and Weigel, 1980, who seek to locate the explanation of nonhuman primate behaviour in mathematical proscriptions. These one-line representations of primate group social behaviour are then treated, perhaps revered, as core or central principles within, not so much nonhuman primate groups as such, but, rather, within groups of human primatologists as they perform and express the discipline. Primatologists have not been sufficiently broadly reflexive to parse apart what is nonhuman primate structure (should it exist) and what is our structuring of nonhuman primate behaviour.

Over the years, a number of us have employed a modified linguistic model to characterize, in a general way, our epistemology. We have employed a combination of surface and deep structures. We have treated the nonhuman primate behaviour as a kind of surface structure and our constructed inferences about pattern in that behaviour as a kind of deep structure. We have treated the latter construction as a code FOR behaviour; i.e., as a proscriptive statement for how "things ought to be."

Reusch and Bateson (1968: 288) stated that "action liberates codified information." What they mean by this notion is that behaviour is an observable surface structure revelation of a rather unobservable deep structure. The deep structure is assumed to be an underlying system or proscriptive code that drives or determines behaviour that is both observed and observable. I shall return to this point throughout the course of this chapter. Primatologists have to be careful that just because we have action or behaviour, we do not necessarily have codes "for behaving" on the part of the primates in the way that Reusch and Bateson suggested in their psychiatric model. We may be able to write general statements that describe the patterns in behaving and, in this sense, we may think that we have codified information or rules OF behaving. But, we must be careful that we do not slip from thinking about patterns OF behaviour to thinking about patterns FOR behaving. We must be *careful*, I think, that we do not slip from discourse about surface structure to discourse that implies deep structure or, rather, we must be careful not to slip from descriptive to proscriptive discourse. While this might be a convenient model or mode of operation in terms of thinking about the behaviour of nonhuman primates, we have reified this notion of a nonhuman primate code and deployed it as an analogue for human culture. Some primatologists have felt that genealogy accounts for large portions of the affiliation data base. (See, for example Chapais, 1983; Grewal, 1980; Hornshaw, 1985a and b; and Kurland, 1977; but see Smuts, Cheney and Seyfarth, 1986 for an extensive review.) One has to be very careful indeed that we do not fool ourselves into thinking that this is a primate version of kinship. Sociohistorical biologists have been tempted to do this epistemological trick of shifting our notion of genealogy to our cultural notion of kinship. The sociobiological construction is often a metaethical construction of behaviour. The implication is, often, that nonhuman primates are conscious or cognizant of their "relations" and that this makes a difference to them. You could pick any article at random from the journals and read that genealogy has become an important explanation. (For example: see Kurland, 1977 for a sociobiological perspective here or Fedigan, 1982, for a less sociobiologlcal perspective.) Our explanations (e.g., age, gender and genealogy) run the risk of being our projections of what we sell as nonhuman primate codes. In other words, we project our code as their code, while we are trying to access some notion of deep structure. What we very clearly have in nonhuman primate social behaviour is still what we could refer to as surface structure. Whether or not primates possess a deep structure is an arguable point. My position on this is that I think that it is possible that there may be such a deep structure, but, more importantly, we have no clean way of accessing this "other species" deep structure. There is no sound epistemology that lets us access "monkey perceptions" in general and monkey perceptions about caging and affiliation in particular. Whatever discussions we construct of monkey perceptions are our enculturated constructions of what we perceive about their behaviours. There are reasonably sound epistemologies that allow us to discuss BEHAVIOUR, but not primate perceptions in the cognitive sense. This is not an assertion that primates have no cognition and no perception - just that, at present, I think that there is no sound, credible epistemology to access these. We simply have to come to terms with our limitations. We have to stop constructing a "not-so-human" primate. We ought to be satisfied that we have access to a very rich surface structure - a richness that would be more easily perceived if we were to stop truncating the view of variability. What I am also suggesting here is that we ought not to deploy "invisible data" in our discourse on the behaviour of nonhuman primates. An area that demonstrates clearly that generalizations of nonhuman primates are greatly influenced by cultural constructions is the research that compares captive and non-captive species with reference to their patterns of sociality. Our cultural notions of incarceration have clouded our ideas about the caged/non-caged issue. In other contexts (Hornshaw, 1985a, b), I have argued that many scientists have regarded the social behaviour of captive primates as "unreal" or as being "nonrepresentative" of what one could see in the "wild" or "natural habitat." There is no question that patterns of sociality differ between and among groups of primates; however, the source of these differences does not lie in the fact that some social groups are captive and others are not. We expect that the source of the difference lies in whether or not the animals are caged. This expectation, test hypothesis, as it were, is a construction of western culture, in general and our notion of confinement in particular. (See Hornshaw, 1985a for an extended discussion.) What makes groups different is the difference in the way that they behave as groups. Rules of behaving, if they exist, are group-generated - not driven by exogenous factors of environment or endogenous factors of a hydraulic notion of genetics. Our explanations of the patterns that we perceive run the risk of being our projection of what we "sell" as the nonhuman primate code. When we rid ourselves of these problems via reflexivity, what we are left with is the behaviour of primates. If we then think of the behaviour as a SYSTEM, then we can talk about information in the system. But, and this is a very important 'but,' we must incorporate into this notion of system, the notion of an existential process; that is, the information distributed does not have an ascertainable purpose. Specifically, primates "sit in proximity" not in order to tell each other that they are "kin sitting in close proximity" nor to tell the primatologist the same thing. What they are doing is either "sitting in proximity" or not.

I first used sociospatial behaviour of nonhuman primates in 1973. As I have discussed before (Hornshaw, 1982) and, as others have stated, the particular strength of a locational method is that locational data are always present (Menzel, 1974; Vogt, 1978). At the time, I was looking for a variable that did not require an interpretive judgement at the point of data collection. Individuals were either present or not, detected or not. An animal was at point X or point Y or not. In addition, a total group locational procedure would not oversample proximal associations or undersample distal associations. I thought that by observing the group as a whole, that I would not be tempted, consciously or unconsciously, to attend too much to certain kinds of individuals or too little to other kinds of individuals. Every individual and all possible associations would stand the same opportunity or likelihood of being represented in the data base. At that time, and subsequently examining other groups (e.g., *M. sylvanus,* the Barbary ape), and re-examining the same population *(M. silenus,* the Lion-tail macaque), over time, I have examined how the three traditionally used variables of age, gender, and genealogy sort with reference to [1] a locational data base, [2] a nearest-neighbour data base, and [3] a grooming data base. I did not know then, nor do I know now, if the monkeys operated according to these principles that I used to explain sociality.

The labels of the variables age, gender and genealogy have been used in two ways: [1] as the descriptor labels for discussing the nonrandom sorting of the affiliation data at the level of "surface structure" and, [2] the same labels but with the additional interpretive level of "deep structure," that is, in this case a primate cognition about age, gender and genealogy. I cannot control how other readers evaluate these labels in my data, but, it would be my preference that the labels remain applied to the level of surface structure only.

Throughout my work I have maintained that what I was really interested in were the processes of social adaptation in both captive and noncaptive groups of primates. The notion of adaptation in a captive environment has received little attention in the literature. This is largely because the notion of adaptation has been derived from examinations and discussions of primates unconfined by humans and meeting the natural opportunities and constraints of an unfenced habitat. Because captivity or management alters the parameters of the opportunities and constraints of the environment, the general notion is that rather than thinking that this "simply" alters the nature of adaptation, that this obviates the processes of adaptation. Our common thinking about the processes of Darwinian adaptation and adjustment has become skewed so that many biologists, in a general sense, think that we cannot come to understand the processes of adaptation in what appears to be a protected habitat. This notion, I think, seriously truncates our possible view and perception of the range of primate adaptability and adaptation.

Our scientific and cultural predispositions incline us to think of adaptation in only a "free-range" or "natural" habitat. Our notions of what constitutes "natural" are culturally constructed. We all too often still perceive that behaviour in captivity is "unreal." We do not yet fully understand the dimensions of the complexity of social behaviour in captivity - or the "free-range" or "wild" habitat, for that matter. We still tend to think that captivity is a homogeneous category; that while architecture and management decisions may vary from one facility to another, and even within the same facility, there is little variation in the complexities of social behaviour. I have demonstrated that while there are differences in the social patterns of groups of primates, of the same species, that the source of these differences did not rest in the fact that some groups were captive or confined and others were not (Hornshaw, 1985b). Because our cultural notions predispose us to thinking that captivity is a homologue for incarceration, we have treated behaviour in captivity as being anything ranging from "different from natural" behaviour to "pathological" behaviour. Another way to put this would be to say that our primatological discourse has characterized the behaviour observed in the "free-range" condition as homogeneous, driven by a few "central proscriptive rules" for the animals to behave by; and that what we have proscribed for the free-range condition, we also proscribe as conditions that "ought not" to occur in captive contexts. We have assumed that all "free-range" behaviour is homogeneous and that all captive behaviour is homogeneous and that the two are qualitatively different from each other.

What is being suggested in the primatological discourse is that captivity ought to present at the level of surface structure, i.e., behaviour, evidence that could be characterized as "unusual" or even nonnormative. The inferred deep structure here is that the habitat as captive or noncaptive drives the surface structural behaviour that we observe. This deployment of captivity as an explanatory variable, actually treated as a constant, though parsimonious, is an extreme version of an environmental paradigm; i.e., that the condition of the environment generates sociality. There is a notion that if we accept the causality of captivity, then we would have some understanding of why groups are the way they are. This is, clearly, a one-word gloss for a very complex set of situations. Captive situations are not homogeneous, and, because this is the case, captivity as a variable should not be treated in a homogenistic fashion. In comparing two groups of *M. sylvanus* (Hornshaw, 1985b), one captive and one not, it became clear that what differences existed between them could not be accounted for by appealing to the constructed notions of the impact of captivity. An exploration of the similarities had a profound impact on the description and depiction of social processes within the two groups. More often than not we tend to concentrate on differences rather than on similarities and it seems important that we ought to pay as much attention to the latter as to the former.

The similarities between these two groups occurred in several dimensions or at several levels. In both the captive and noncaptive groups there was evidence of what has been called "preferred" associations and that genealogical connections or affiliations accounted for much of this association. The smallest inter-individual distances, in both groups, were in the class of mother-infant associations. I have stated that monkeys in the same "matriline" were significantly closer together than were monkeys that were not. What I intended by the use of this term "matriline" was the denotation that these were animals that, to use a human kinship term, shared the same "grandmother." Unfortunately the use of the term "matriline" has a connotation that goes well beyond this limited statement of consanguineal relationship. The data from both groups demonstrated that associations do not work on the basis of simple syllogism, i.e., just because monkey A is consistently close to monkey B and monkey C, it does not necessarily follow that monkeys B and C also exhibit "proximal association." In both groups, it was also clear that each animal expressed a distribution of associations ranging from very proximal to very distal. No animal in either group expressed exclusively proximal or exclusively distal associations. In neither group was there a kind of inside-outside or centre-periphery organization, either related to any environmental feature or to any particular individual. To state it more bluntly: expressing distal associations with one or even several animals did not preclude proximal associations in general or with specific individuals.

There were differences between the two groups. The zoo group, as a whole, expressed *larger* inter-individual distances than did the non-zoo group. Although the latter had access to a considerably larger habitat, it was significantly closer together than was the zoo group. In both groups, the range of inter-individual distances was similar, i.e., the means were different but the ranges were similar. Having less available space did not compress the inter-individual distance affiliation. At the time, I explained this difference by examining the similarities in the two groups. Members of both groups, recall, appeared to associate on the basis of matriline association. In the zoo group the depth of the genealogy, at the time of study, was two generations while the depth in the non-zoo group was three generations. It looked as though having a third generation intensified the already existing pattern of affiliation or, stated more simply, that having another generation of "matrikin" literally added more matrikin with which to affiliate. Basically, increasing the cell size strengthened the statistical association. The point that I am making here is that the statistical reality may obscure or not reflect a behavioural reality. I will return to this point somewhat later in the chapter.

Our research procedures have been developed, amongst other things, as a ritual process that because of their reification have led us increasingly to homogenize our data and our construction of appropriate research questions. We think, sometimes, that what we are interested in is some kind of primate code for behaving. If I do exactly the same thing each time, I make my data more comparable from one data set to another. At least, I am holding myself, the transformer of the data, constant. There is associated with this a *colloquial* notion that if I hold something constant, I can, essentially, ignore its effects in the transformation process. If I make myself homogeneous with respect to my behaviour of "recording data," then this somehow increases my validity and the putative validity of the data base. I am not recommending a procedural "free-for-all." I am simply examining our commitment to homogeneity and stasis. If I collect "exactly the same thing" from one situation to another - like who grooms whom or who sits next to whom through time, then I might come to think that what appears in my notes *is* exactly the same thing. What I have operationally decided is the same thing, appears and reappears in my field record. This leads to a potentially false sense of trust based on a notion that what has been defined as "looking the same" is, in fact, the same.

One of the things that my own data have demonstrated has been that what is defined as looking the same and, therefore, recorded as being the same, is not. For example, the value of the proximity variable "full body contact" was, and is, an easily observable and easily recordable operationally defined behaviour. It can be recorded as yes/no without any "subjective" appraisal for further qualification such as the assessment of the presence/absence or even character of "tension." So we treat "full body contact" as an operational constant on our part but make the epistemological error of thinking that this is a constant on the part of the primates. This, though being procedurally clean, is anthropomorphic in a clear sense. Having examined two groups of the same species maintained at the same facility, *M. silenus* in this case, it has become clear that what I used as a discrete easily observable and unambiguously defined behaviour may not have been a homogeneous class of behaviours. If I were, then, to use this behaviour as a surface index of some primate deep structure, then I would be in serious trouble, because what I have treated as being homogeneous is not. What homogeneity exists, does so, I suspect, inside the operational definition. This is the danger of thinking traditionally in the sense of lock-and-key definitions. We have been complacent about our operational definitions and our operational notions of what constitutes "units of behaviour" - those units that we record in the field. In some of my recent work, re-examining groups of *M. silenus* maintained at the same facility, it has become evident that, for example, "full body contact," though topographically the same at one level, may not mean the same thing in different groups or even the same group at different sociohistorical times.

If, as I have stated, I am really interested in the processes of *social adaptation* in captive groups of primates, then, the first thing that I have to do is "allow the groups to speak for themselves," at least as much as possible. Obviously, this is a metaphor for saying that I should be careful not to superimpose my notions onto the primates. To continue with the example just used, i.e., the category of "full body contact," this behaviour of "full body contact" was not the same behaviour across groups, but, it was recorded as though it were. The procedural homogeneity tended to *render* the groups more similar than they, perhaps, were. Part of this project involved a proximity data base. The proximity referred to here was a nearest neighbour analysis rather than a locational data base (as I had originally collected in 1972). In reality, the proximity data is a truncated version of a locational data base. Whereas the latter generated a sequential plot of every individual's location through time with reference to [1] the available habitat and [2] every other animal in the group; the proximity data pays no attention to the place of individuals with reference to architecture or habitat; nor does it record the social location of each animal with reference to every other animal. It tends to sample proximal associations rather than the range of associations. Both locational data and proximity data were chosen because they are discrete, easily observable behaviours, that do not require a judgement at the point of data collection. Because proximity data as nearest neighbour data samples 'who is close'; it does not sample 'who is far' and, therefore renders the appearance of the group as a whole, more proximal and more homogeneous than it perhaps is. It effectively ignores or truncates the recorded range of social positions. I want to develop this notion of 'position.' It has been treated as a procedural constant. The notion of 'position' carries with it the notion of stasis. The commitment to static procedure has been accompanied with a commitment to static interpretation. A good example here would be Altmann's 1974 'milestone' paper. While this particular source is now somewhat dated, it has been used *extensively* as a proscriptive model for the production of both primatological research and primatological discourse. (There are others; see, for example, Altmann, 1980; Bramblett, 1976; Hinde and Atkinson, 1970; Hinde and Spencer-Booth, 1967; Hrdy and Hausfater, 1984; and Kurland and Gaulin, 1984). Over the years, in primato-logical discourse, we have examined the influence of three variables: age, gender and genealogy or line relationships with reference to a variety of data bases. In this context, however, even though we have used the term or notion of "variable" in our scientific discourse, we have tended to treat that which is actually variable as though it were a constant. Once we have demonstrated that a particular association in a data set of the form: "genealogy accounts for most of the variance," then we become satisfied that we have demonstrated the importance of this particular variable (or any other possible one) in a particular group of a particular species in a particular context. And because we have been committed to metaethical notions of both stasis and homogeneity, we often do not expect groups of nonhuman primates to vary with respect to our construction of how they "ought to behave."

Groups of conspecific primates maintained at the same facility under the same management decisions do not necessarily express the same patterns of sociality. This is another way of pointing out that captivity is not a homogeneous variable or class of situations. When we examine groups looking for scientific one-line statements (viz: "genealogy accounts for large portions of the 'affiliation data'), we often commit ourselves to a static view of the patterns of sociality in a particular group. Groups change over time with reference to their behaviour and, ultimately, our view of that behaviour ought to change as well. At the Assiniboine Park Zoo, one of the Lion-tail groups (GH) did not show a "statistically significant" association between affiliation as proximity and "line membership." This was interesting because in past studies (see Hornshaw, 1985a) of these groups, this statistical association was demonstrated. There are several points to be made here. The first is that different groups of the same species maintained at the same facility do not necessarily express the same homogeneous patterns of behaving just because they are maintained at the same facility. Secondly, just because a group appears to behave in one way at one time, there is no reason to think that it will do so at another time. Behaviour is embedded in groups, who, regardless of species and habitat can and do differ with regard to their behaviour patterns.

There is another possible interpretation that we ought to consider seriously here. While it is possible that different groups of the same species maintained at the same facility differ with respect to what might be used to account for patterns of affiliation, and, while it is possible that any particular group of a particular species can change through time with respect to what principle appears to account for patterns of affiliation and distribution; it is also possible that we were mistaken from the beginning with regard to our interpretation. To be more specific: at time X, the silenus group that I am currently discussing, demonstrated the statistical association between affiliation and genealogy, but, on subsequent examination, at another time, did not. One must entertain the theoretical possibility that this group never organized themselves according to the central principle of genealogy - that this appearance was the result of a statistical sort coupled with interpretation. One must try not to confuse a statistical reality with a behavioural reality. It is also quite possible that the internal parameters and dynamics changed through time and that these changes could account for the changes in the patterns of behaviour observed. Over the years, genealogy has become an important factor that has been deployed in explanations of primate social behaviour. Sociobiologists like Kurland (1977) and non-sociobiologists like Fedigan (1982) are keen to demonstrate the apparent relevance of line membership, genealogy and kinship. I have also demonstrated this relationship in a number of my studies on both *Macaca sylvanus* (the Barbary "ape") and *M. silenus* (see Hornshaw, 1985a and 1985b). Each group, sociobiologists and non-sociobiologists, has different theoretical and epistemological reasons for doing so, but we have all become wedded to the notion of the importance of this variable; so much so that it has been treated as a constant. In a recent *M. silenus* study, one group demonstrated a statistically significant association between proximity and 'line membership' and the other group did not. If I "pooled" the data from both groups, then there was a statistically significant association. This would "salvage" the genealogy constant but would hide the variability and variation between the two groups.

There is a reason not to pool the data here. The fact that two groups of the same species, maintained at the same facility, with a shared social history, demonstrate different patterns of affiliation speaks to the issue of "local traditions." It might be that in spite of shared species, shared management decisions and a shared social history, that different groups, as a result of processes uniquely internal to each group, deploy themselves differently. As I have stated previously, one of the important issues in primatology has been the extended discussion of the importance of "kinship" in primate social organization. I have just discussed my own data in this regard. What is important now is how we deal with this data, i.e., that different groups of the same species maintained at the same facility, with a shared social history demonstrate different patterns of affiliation. Sociobiologists might become concerned because, from that general perspective, one would expect all groups of the same species to be homogeneous with respect to the putative importance of genealogy and kinship, i.e., that this would be a characteristic of the species repertoire. Sociobiologists could salvage their understanding by invoking phenotypic variation or variation at the level of surface structure. But, for them to do so would be a risky epistemological endeavour because they must use this "surface behaviour" as both an indicator and validator of the inferred and constructed deep structure of genetics, in this case. Non-sociobiologists might be tempted to rejoice because they could interpret this variability in the effective expression of the relationship as being evidence for different local social traditions, that, in some cases, make it appear that genealogy and affiliation sort significantly together - that different groups express unique internal social processes. Care must be taken here as well, so that we do not use behaviour as surface structure as a window to an equally unobservable deep structure of collective cognition. It does not end here, because all that sociobiologists would have to do is embed the notion of tradition in general and local traditions in particular back into the notion of a species genetic repertoire. Then they could say that the different local social traditions reflected different local gene pools, as Wilson (1978) attempted with Burton and Bick's (1972) notion of Tradition Drift. So, it looks like we might have come back to the same point, i.e., that one can interpret one and the same data from two opposing perspectives, a sociobiological one and a non-sociobiological one. Is there a way out of this dilemma of the binary opposition in interpretation?

Examining the same groups through time might shed some light on this problem. If, at time X, a group or groups demonstrated, in this case, a statistically significant association between genealogy and affiliation, and then did not at time Y, it would be difficult to argue that this change in patterns of sociality reflected shifts in the genome, particularly if the individuals that were present at time X were also present, with additional members, at time Y. One might be more inclined to perceive the time successive variation in the putative relationship of this variable with the patterns of affiliation to be a reflection of social processes uniquely internal to the group. This notion might be strengthened if a second group, again of the same species, continued to demonstrate this relationship time successively. This is, at least, what my data has led me to think. Earlier work on this group (Hornshaw, 1974) had demonstrated that the smallest interindividual distances were found in the class of mother-infant associations. These mother-infant distances were not homogeneous, however. There was considerable variation in the distances expressed by the three pairs of mothers and respective infants. Likewise, there was variation in the distances of the three possible consort associations. The individual that expressed the smallest interindividual distance with the adult male was also the female that expressed the largest interindividual distance with her own infant. In short, associations do not work by syllogistical principles. What occurs in one set of associations, like consort associations, reverberates into other sets of associations, for example, mother-infant associations. Another concern that has emerged from recent *M. silenus* material is the lack of quantitative concordance between grooming measures of affiliation and the proximity measures of affiliation. At some level, this must be surprising for the primatologist. Both grooming and proximity have been used as surface structure indices for the deep structure of affiliation or social organization. Indeed, they have been used interchangeably. Intuitively, from a *human* point of view, parsimony would suggest that grooming and proximity would generate similar data or, would, ultimately, access the same social processes, if for no other reason, than the mechanical fact that grooming necessitates proximity at least for the duration of the groom. What is at issue here is that we may have assumed that affiliation is a single homogeneous process. My most recent *M. silenus* material demonstrates that there is no relationship between the grooming and nearest-neighbour data. It is interesting to think that grooming and proximity may subsume quite different aspects or forms of affiliation. What is clear is that grooming and proximity not only generate quite different data; they mediate different social processes. Once again, things which look the same or look as though they ought to be the same, from an interpretive perspective, are, clearly, not the same either in topography or potential meaning.

Conclusion

We, as primatologists have become so concerned with *meaning*, that is, assessing what the behaviour means, that we have treated all behaviour emitted by primates as carrying meaning. It is possible that not all behaviour is meaningful to the primates and to the primatologists. From a human perspective, many of us, at one time or another, have made disparaging remarks about psychiatrists and how they *tend*, we think, to overinterpret behaviour. They are persistently asking what every X or Y behaviour means and, to us, they often appear to assume that every X or Y behaviour has meaning, carries meaning or is linked to a system of meaning. We have all heard the jokes that caricature psychoanalysts that say: "you can tell a great deal about a person (i.e., a deep structure) by the way that they, for example, open a banana, drive a car or dress" to mention but a few surface structure possibilities. Psychoanalysis seems to assume that behaviour is a surface representation or 'acting out,' for a deeper structure of meaning. This may be the case for humans in an enculturated system; but, it is not clear if this model applies to or for nonhuman primates and their behaviour. I think that we might want to entertain the possibility that not all behaviour on the part of nonhuman primates carries *meaning* either for the primates or the primatologists. We ought to make a serious beginning to entertain analyses of "noise" in the information system; or, at least begin to see that not all behaviour be required, by our construction, to carry meaning. We must guard against thinking that every primate behaviour is meaningful in an interpretive sense. Perhaps we, like our caricature of psychiatrists, have overinterpreted some primate behaviour. It may be that the deployment of, for example, "kinship" may be one of these overinterpretations. Whether we use behaviour as a vehicle to access a deep structure of genetics or a deep structure of cognition, in both cases, we may be overinterpreting [1] the data and [2] whether the *behaviour*, as data, is a surface structure reading of a deep structure at all. This point allows me to return once again to a distinction between patterns OF behaviour and rules OF behaviour as contrasted with rules FOR behaviour. It is clear, I think, that we can think of primate behaviour as a kind of surface structure we can, quite reasonably, write generalized statements that characterize that behaviour; and, in this sense characterize patterns or 'rules' (as generalized statements) OF behaving. We must, I think, be careful that we do not engage in a kind of intellectual "slippage" that permits us to think that we have clean, credible and valid access to a deep structure or code FOR behaving.

In discussing Prigogine's epistemology, Maruyama (1978) elucidates a number of critical points relevant to this discussion. Prigogine and Stengers (1984) and Maruyama (1978) point out that:

> traditional science operates on the assumption that the most probable state of the universe or of an isolated system is random distribution of independent events, each having its own probability. In such a universe less probable states of nonrandom distribution tend to change to states of more random *homogeneous* distribution. The universe converges to homogeneity (Maruyama, 1978: Emphasis mine).

It is this working assumption of homogeneity that I have been examining in this chapter. In traditional cartesian science, research has been designed to search out differences but in a way that asserts 'sameness'. Traditionally, even though we have searched for differences, this search for difference has been based on the assumption that "similar conditions produce similar results" (Maruyama, 1978: 453) but this assumption is believed to be validated retrospectively, i.e., if the results, for example, of two bodies of data are similar, we have inferred that the causes or conditions giving rise to these results are similar and, if the results from two bodies of data are dissimilar, we have inferred that the causes are dissimilar. Similar conditions may lead to dissimilar results and, quite possibly, dissimilar conditions may lead to similar results. The search for dissimilar results, for example, in behaviour has really been the search for dissimilar causes or dissimilar "deep structures"; in this specific case, the deep structure of genetics, environment or even the construct of cognition. In the exploration of autonomous local group social traditions or constellations of behaviour, I have attempted to incorporate a heterogenistic (in the Prigoginian sense) epistemology, rather than a homogenistic one. As Maruyama (1978) has pointed out, there has been an unfortunate intellectual convergence between the two notions of homogeneity and 'peace.' Our ideological construction of "natural nature" is the construction that it is "peaceful." Differences do not in and of themselves generate or reflect conflict, and similarities in and of themselves do not generate or reflect a peaceful state of affairs. Nature and biology are not dedicated to homogeneity or "peacefulness"; they are simply heterogenistic, constantly in dynamic flux and differentiation. As Prigogine and Stengers (1984) and Maruyama (1978) have discussed, we have tended to interpret differences as reflecting notions of superiority and inferiority. Such has certainly been the case in the primatological literature that counterpoints noncaptive and captive sociality. Where differences have been observed, it has generally been concluded that the captive environment is the inferior environment. Part of this chapter has been an exploration of the fallaciousness of the underlying assumption in this paradigm. Groups of nonhuman primates express both similar and different social processes but the source of these differences and similarities lies in the CONTEXT of each group's sociality, not necessarily in each group's genome or habitat. This notion of group sociality or local group traditions is heterogenistic not homogenistic. Similar conditions are not necessarily associated with similar results and, dissimilar conditions are not always associated with dissimilar results. When groups change through time with respect to group sociality they do so not necessarily because the gene pool or habitat have changed but because their local group social processes are themselves heterogenistic.

CHAPTER SIX

An Alternative View: Behaviour As A Multi-Causal Strategy For Survival

James D. Paterson

Behaviour, particularly social behaviour as studied in primatology, is an activity which can only be reported to others as the processed and interpreted perceptions of an observer. Candland and Kyes (1986: 26) confirm this perspective and note that we are "...unlikely to make substantial progress in understanding animal life until we have first understood the nature of the categories that we impose upon our perceptions of other species." According to Asquith (1984, 1986) these categories are inevitably related to anthropomorphic perceptions. In consequence, behavioural events and states are frequently reported as the results of internalized and sometimes unperceived operations on the part of the observer, and may or may not have a high level of correspondence with the activities as they might be reported by the subject primate itself. This process of observational abstraction may lead to an overreliance on a single organizing principle or causal model. It is my intention to argue for a multi-causal synthetic which links social and individual behaviour into an interactive relationship involving the genes, physiology, learning capacities, cognitive functions, self-awareness, and the external evolutionary selective pressures. I will then discuss some aspects of thermoregulatory physiology and present data and conclusions from a field study on behavioural thermoregulation to partially illustrate the functioning of the synthetic model.

The model makes use of the capacity for flexible or plastic response at the physiological level, and a similar but more extensive flexibility at the behaviour level. If a selective pressure begins to affect a species, the first response by the individual organisms may be to accommodate the forces by a behavioural modification; if that is insufficient, a physiological change may accommodate the situation, and only if the selective pressure is sufficiently large, or of a form which bypasses these mechanisms, may it have an actual selective effect upon the structure of the population's gene pool. While this model has primarily been viewed as an evolutionary scheme, it also is a model of behaviour generation. Only this latter aspect will be emphasized here. Additionally, the model incorporates consideration of the roles of learned behaviour, cognitive thought processes, and individual self-awareness as layers within the proximate causality of an organism's observed behavioural pattern.

Causality and Interaction Effects Relevant to Modeling Behaviour

Within the realm of the behaviour of biological organisms, the level of complexity is extremely high. The result of this complexity is that mathematical arguments are less and less absolute, probability becomes a major factor in all analyses, and the mono-causal argument becomes less useful as an explanation of the *observations*. Certainly no single, simple experiment can confirm a hypothesis, although it may contribute significantly towards the affirmation of one. A point to note for the later discussion is that a perfect correlation between two biological or behavioural phenomena is very rarely observed. When such an occurrence takes place, it does suggest a mono-causal condition; however, since most correlations are far lower, these tend to strongly suggest that multiple causes are involved. Therefore, in my own thinking about behaviour, the mono-causal argument has been downgraded as being less relevant to the analysis, and a multi-causal perspective is most frequently invoked. It is within this multi-causal environment that the synthesized model of behaviour is discussed.

The problem of cumulative interactive effects, otherwise labelled as synergism, is one with which all biologists must wrestle. In a recent paper, Mason summarizes the history of attitudes towards the study of behaviour as follows:

> Attitudes toward animal behavior have changed much since the time, a few years ago, when scientists somewhat self-consciously proclaimed that their concerns were restricted to 'objective' behavior. The implication was that an adequate and complete science of behavior could be based solely on what was observed. To be sure, it was widely recognized at the time that a satisfactory general account of behavior required the assumption that various covert processes and hypothetical mechanisms played an essential role in the causal network that determines what an organism actually does; hence the familiar and time-worn concepts of learning, memory, perception, decision-making, needs, motives, emotions and the like. At the same time, the prevailing attitude towards such hypothetical entities was cautious and conservative. In the name of parsimony and objectivity, the formidable complexities surrounding the ontological and theoretical status of hypothetical processes were avoided by the expedients of explicit operational definitions and a data language that, ideally, was stripped of all mentalistic connotations. (Mason, 1986: 3)

Mason continues to discuss the potentiality for political behaviour in primate groups, and comes to the conclusion that political aspirations and activities are thoroughly embedded in the lifestyle of all social primates. He proceeds to state "...it is clear that infanticide, dominance, the formation of coalitions, and many other features of primate social life are not the expressions of simple traits but the outcomes of complex interactions among individuals and situations" (Mason, 1986: 10). In this perspective, multiple causality is an explicit component.

It is the common situation in behavioural ecology[1] that any analysis of an observational data set will eventually yield a "proximate causality" which is reducible to a number of contributing causal factors having relatively immediate effects, and an "ultimate causality" which involves the evolutionary significances of the data. The usual state of affairs is that these contributing factors account for only part of the total proximate causality, with a substantial portion being due to "unknown factors." The mathematical tools for dealing with these multiple factor models are moderately complex, but still fall short of being completely adequate. While they are able to partition out the multiple factors, and attribute some reliable portion of causal responsibility to each, they are not yet able to deal with the how and why of the interactive effects between the factors, nor provide any direction in which to search for the "unknowns." The only tool which seems to be oriented towards some aspects of these problems is the LISREL system of Joreskog, Sorbom, and Magidam (1979), which performs correlation calculations between the "residual" values of regressions which are not otherwise incorporated into the analysis procedure. Some of these statistical tools (various correlation and regression models but not LISREL) are used in the analysis presented in the second part of this paper.

A Buffered Model of Behaviour

The model of behaviour proposed is depicted as Figure 1.

Figure 1: A Model of Behaviour Generation

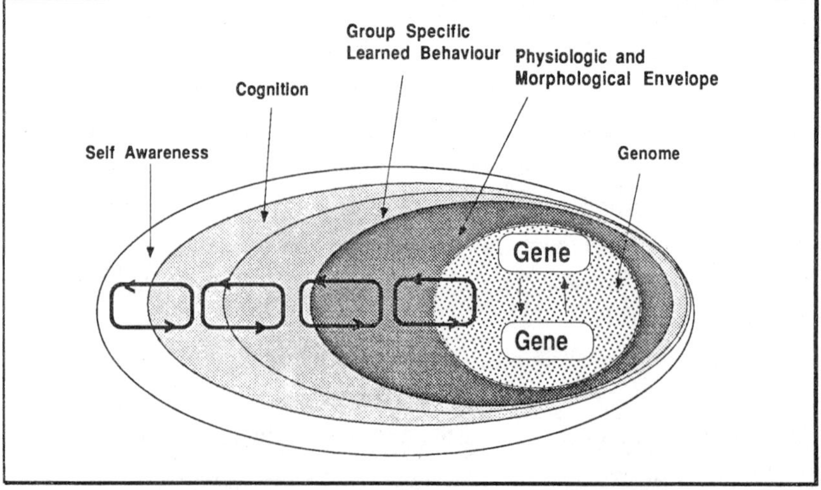

The model is composed of a number of successive shells, each representing a different aspect of causality, with the genetic constitution of the organism at the innermost level. Each shell forms a "buffer" between the adjacent shells. By the term "buffer" is meant a layer which absorbs, cushions, or redirects the pressures applied by environment on the organism. At the same time, the model expresses the perspective that the interacting behavioural shell is influenced, generated, and controlled to some extent by the deeper layers through homeostatic mechanisms, and feeds back some influence to that deeper layer. While it would be theoretically possible for a feedback loop to extend directly across two levels, thereby circumventing the buffering effect of the sandwiched layer, such an increase in complexity is not seen as necessary, since the chaining of feedback can accomplish the same effect without the additional construction. The model may be simplified by noting that the three layers labelled as self-awareness, cognition, and group-specific learned behaviour may be collected as a single layer identified merely as the behavioural shell, as depicted in Figure 2.

Figure 2: Three Level Model of Behaviour Generation

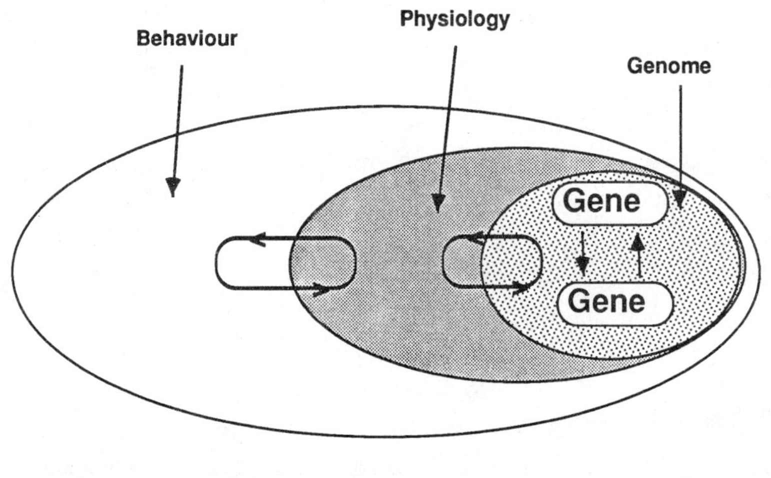

As field observers we can only examine this outermost of protective shells, the organism's behavioural activity, and must infer the functioning of the various layers within. To examine the inner layers more directly, it is necessary to rely upon disciplines which are not restricted to purely observational methods. For the purposes of field observation and this discussion, behaviour may be viewed as the organism's perceivable interface with its environment (in line with Goodall's description of behaviour as being anything the animal does). The so-called 'behaviourist' school, which followed the tenets of B. F. Skinner (e.g., Skinner, 1953) felt that this was the only level at which the analysis could or should be carried out, that behaviour could only be explained in terms of its consequences. Most behaviourists dismissed the idea of cognition and consciousness in animals without consideration (Rumbaugh, 1986). Modern sociobiology, in its most rigid formulations, takes an opposite extreme by essentially attributing almost all behaviour to some form of genetic basis (Wilson, 1975). As the model intimates, the synthesized perspective as reflected here concurs with the contemporary view that behavioural causality is neither behaviourist nor sociobiologist, nor rigid in origin.

Behaviour as Epiphenomenon

In one sense, an organism's behaviour is only what the observer "sees and records" (Altmann, 1974), unlike Goodall's definition of behaviour as "anything that the animal does," and is therefore an imposition of arbitrary form on what the organism is doing (Mason, 1986; Candland and Kyes, 1986). That is a major problem of observation, but is quite separate from the problem of behavioural causality, what generates the behaviour? Since field observers cannot directly observe, but can only infer the internal operations of the behaviour-generating mechanism, they can only treat behaviour as an epiphenomenon. By this I mean that observers record the form and pattern of behaviour and treat it as if it was unbiased, unselected, objective data. They do not normally concern themselves, at this level of analysis, with the problem of causality. Under these conditions, the study of behaviour can be treated as an end in itself, and behaviour itself becomes an epiphenomenon, an arbitrary categorization created and labelled by the observer, existing without established linkages to whatever mechanisms of causality are operating within the organism. To probe further into the internal workings of the behavioural generating system, other disciplines must be consulted, such as the review of perspectives on animal cognition and consciousness provided by Roitblatt, Bever, and Terrace (1984).

Details of the Model Layers

A consideration of the importance or impact of each of the layerings of this behavioural model to the overall observable interface is the major part of the next section of the argument.

Level 1 - Self Awareness, Cognition, Learning and Group Tradition

Level 1a - Self Awareness

Self-awareness or consciousness is placed at the outermost layer of this behavioural model, in part because it may be the latest elaboration in the evolution of the neurological control system (which Gallup, 1982 refers to as a "cybernetic" or self-steering/self-organizing system). It has also been the least consistently demonstrable capacity among the primates. 'Self-awareness' implies that the organism recognizes itself as being a separate entity, different from all others, and thus has some "directedness of behaviour" which Singh (1986) and Shafton (1976) considered to be above and beyond the functions of cognition. This self-awareness may result in the behaviour patterns which have often been labelled as "selfish" when they are examined in the evolutionary milieu. The utilization of the mirror test and the mirror and marking experiments in primate studies (Gallup, 1977; Platt and Thompson, 1985) have shown that the great apes can be regarded as self-aware, but that the capacity is spottily distributed in monkeys, with some members evaluated as self-aware and others unaware, even within the same troop. However, it must also be noted that Bramblett (1989) has criticized the research design of Gallup as being fatally flawed. In an evolutionary sense the individual organism which is self-aware may be perceived as expending a greater portion of its energy resources on self-protection, apparently to act in a more "selfish" fashion, to use Dawkins' term (Dawkins, 1976), and thus increase its probability of survival, status enhancement, and eventual reproduction (Reynolds, 1986). While Dawkins attributes selfish actions to a mechanism based upon the genes, this model suggests that selfish activity may be correlated with self-awareness, residing at the outer buffer layer rather than the innermost, that is, it may be at the proximate level of causality rather than at the ultimate.

Level 1b - Cognition

Cognition, here defined as the capacity to think, to perceive, recognize and understand, is involved with the ability to selectively accept data input and to calculate proximate outcomes. Making decisions based on those calculations is an ability which is widespread amongst animals. While some researchers may deny that a particular set of organisms are capable of cognitive activity, others, myself among them, regard it as a feature of nearly every organized brain structure. Anderson (1983: 20) however, by including "...a host of complex skills required for such functions as perception, memory, language use, reasoning, problem solving, and learning" effectively restricts cognition to *Homo sapiens.* Burton (1984), however, has presented arguments that indicate - from observation of behaviour - that cognitive capabilities must be acknowledged in non-human primates. Cognition, as used here, is more than the capacity for direct stimulus and response (Singh, 1986; Reynolds, 1986; Asquith, 1986), but it does not require the addition of language to the matrix. If cognition can be equated with thinking, then clearly, amongst its activities are the processes of evaluation, of assessment of risk, and the implicit application of a "mini-max" (minimum expenditure for maximum gain) or an "optimax" (optimum expenditure to ensure maximum gain) calculation, before action is initiated (also see Reynolds, 1986; Anderson, 1983, for views of the social utility of cognition). The cognitive process is coincident with what many would consider as "intelligent decision making," or "conscious social action" (Reynolds, 1976; Harre and Secord, 1972).

Level 1c - Learning and Group Tradition

That higher vertebrates, and even some lower forms, display learning and group traditional behaviour patterns is a widespread perspective in evolutionary biology. It is impossible to advocate a concept of "tradition" without accepting the capacity for learning as a precursor mechanism. That learning is an important aspect of behaviour and of great significance in evolution has been accepted in the literature for a very long time. In 1971 Braestrup noted its evolutionary importance in this fashion:

> ..the evolutionary role of learning, the adaptive modification of behaviour, is especially prominent in higher vertebrates. As emphasized by Mayr (1960, 1963,1970) behaviour has a leading part in all kinds of evolutionary shifts because in birds and especially in mammals behaviour is very plastic. (Braestrup, 1971: 90)

As far back as the beginning of this century, Baldwin (1902: 180) spoke of the relationship between learning and tradition as "a constant adjustment, as the conditions of life may demand, between the intelligent actions embodied in tradition, and the instinctive actions." This perspective was recently resurrected and discussed in a series of papers at the Tenth Congress of the International Primatological Society at Nairobi in 1984. Cognitive and learned aspects of foraging and behaviour were discussed for *Saimiri* (squirrel monkey) and *Callicebus* (the titi monkey) (Andrews, 1986), *Papio hamadryas* (the hamadryas baboon) (Sigg, 1986), and *Alouatta* (the howler monkey) (Whitehead, 1986). While simple learning may be adequate to understand the immediate responses by individuals, tradition implies that the learning takes place within a gregarious group with the learning of younger individuals being constrained by older group members. In this way, tradition builds upon the experiences of the previous generations and represents a cumulative learning set for the group.

The well-known cases of traditional differences in behaviour between groups of vervet monkeys (*C. aethiops)* (Gartlan, 1968), Japanese macaques *(Macaca fuscata)* (Miyadi, 1964; Tsumori, 1967; etc.), bonnet macaques *(M. radiata)* (Singh, 1986; Singh and Prakash, 1981), barbary macaques *(M. sylvanus)* (Burton, 1972a, b), and savanna baboons *(Papio anubis)* (Paterson, 1973, 1976; Altmann, 1974; Hausfater, 1976, 1982; Nagel, 1971; Packer, 1984; and many others) make it empirically obvious that the maintenance of a group-specific learned behaviour pattern is supportable for many species, if not most. It is a necessary component in the model. Since traditional differences may be found between groups of the same species, it becomes obvious that a mechanism for change in tradition is required to account for it. This can be found in the model of "tradition drift" proposed by Burton (1972) as a *cultural process*, but altered by Wilson (1975) to serve as a genetic tenet of sociobiology. Tradition drift suggested that the act of learning traditions in each generation results in changes of pattern from earlier forms due to behavioural variations in performance, observation and memorization. Such a mechanism is reminiscent of the Sewall Wright Effect in genetics, and presumably, like genetic drift, would convert to either a stabilizing function or a faster pattern of change should selective pressures take a hand in the process. A more recent incarnation of this pattern of argument is the discussion of "memes" or "units of ideas" which Dawkins (1987) uses to extend his model of Darwinian evolution.

Level 2 - Physiology and Morphology

At level 2, the model moves into the realm of the chemical world. The physiological processes and the morphology of an organism are directly connected to the operation of the genome (John and Miklos, 1988), and in a large percentage of instances the genes appear to have direct control of various aspects of physiology and morphology. However, it has been long recognized that these structural aspects are subject to flexible or plastic expression. Whether this is termed epigenesis or variable expression of a gene is irrelevant to the discussion. This flexibility is at the basis of the phenomenon which Waddington (1957) labeled as "canalization," and it underlies the common genetic terms "expressivity" and "penetrance." The term 'expressivity' refers to the degree of expression of some variant phenotypic character. This means that a gene-controlled, and observable, characteristic can vary substantially in form and pattern. "Penetrance" describes the proportion of a given set of zygotes which develops into adults exhibiting a particular and recognizable phenotype. This means that not all the organisms which have a particular gene will develop the particular phenotypic characteristic. These latter two terms were first set out by Timofeeff-Ressonsky in 1927 (Waddington, 1957). Canalization, according to Gould (1977), can be seen as the set of phylogenetically derived ontogenetic constraints which direct the development of individual organisms. Waddington graphically viewed this process as a slightly lumpy channel, occasionally containing lateral divergent tracks on the walls, down which the ontogenetic development rolled like a ball. If unusual conditions force the ball of ontogeny upwards onto the wall of the channel and into one of the divergent tracks, significantly distinctive physiology or morphology could arise, but if ontogeny remained in the main channel, which constitutes the line of least resistance, then ontogeny would proceed to a normative conclusion.

In developmental and ecological studies, there are numerous instances of variation in morphology and physiological processes which are related to the supplies of necessary chemical resources for development (Kleiber, 1961; Brody, 1945). In these cases, inadequate, normal, or excessive general nutrition supply can result in distinct and recognizable differences in morphological structure, or in physiological variations in the metabolic pathways. Changes in structure or metabolic functioning are particularly associated with inadequate supplies of trace elements and vitamin precursor compounds (Brody, 1945; Richard, 1985). Similarly, variations in structure and metabolism can be attributed to an adjustment to the environmental conditions under which development takes place (Connell, 1984). For detailed consideration of these aspects in humans, the recent volume by Harrison, Tanner, Pilbeam, and Baker (1988) is an excellent source. Nutrition and disease patterns may be the major factor influencing development (Harrison, et al., 1988), but environmental acclimation also has a substantial effect, as will be shown below. That morphology influences behaviour is superficially obvious. The structure of an organism directly limits the ways in which it can behave. It would be ludicrous to expect a mole to brachiate, a gibbon to tunnel, or as Burton has noted, "monkeys cannot fly; dogs cannot grasp" (1972: 29). The morphology of each animal is quite unsuited to the behavioural activity of the others. Morphology thus sets a limitation range or envelope of possibility within which the organism may operate freely, but beyond which behaviours become physically impossible. To modify a terminology from Waddington, it may be said that morphology "canalizes" the pattern of behaviour by making it difficult or impossible to perform behaviours that are near or at the limits of an organism's structural capability.

Level 3 - The Genome

This area of the model is concerned with the mechanisms which have often been claimed to be "hard-wired" or structurally incorporated into the organism's central nervous system (Lorenz, 1965), in particular the limbic system of the midbrain, and with the apparent set of "species-specific fixed-action patterns" which are the observable results. While it is impossible for the genome to produce formed behaviours by itself, there are many instances in which Schneirla (1956, 1957, 1966) and Lehrman (1970) would agree that the influence of the genome is strongly represented in the observable outcome. The interaction of the genome with the environment, through the buffers proposed in the model, agrees with the perceptions of Schneirla and Lehrman as to the importance of epigenetic factors in behavioural causality. However it remains necessary to accord the genome an important place in the causality of behaviour. The genome is the seat of the encoded information which serves as the plan for the ontogenetic development of the organism. Thus, indirectly, where neural structures dictate a particular response to a particular stimulus, the genome can be said to be responsible for the action. It is unwise, however, to view the genome solely as a set of plans for development, for as John and Miklos (1988) point out, the genome is constantly active, genes are jumping from chromosome to chromosome, multiple copies of genes are proliferating, and the suite of activities within the genome of each cell resembles an evolutionary paradigm in miniature.

In this multilayered model an attempt to rationalize and integrate the different perspectives on behaviour is accommodated through the recognition that for some aspects of the perceptual behaviour shell, the overlaid zones are very shallow, in other areas very deep. It is the areas of shallow overlay that the most direct effects of the genes are seen on behavioural patterns, though it must be recognized that the link from gene to behaviour remains an indirect one with a substantial shell of physiological processes between them. The sociobiology and behaviourist perspectives have expended large amounts of their energies on relatively narrow ranges of the behavioural envelope, those which are relatively closely influenced by the genes for the former, and in the areas where the layers are very thick for the latter. Consequently they may be perceived to have over-enthusiastically extended their findings to all behaviour.

Multiple Feedback Loops

The model in Figure 1 also displays a set of "feedback loops" which provide the opportunity for one shell to influence another. These feedback mechanisms will for the most part be of the negative feedback or stabilizing form. This implies that a change in one shell will elevate or lower responses in another shell in order to maintain fluctuation within a controlled range. The two shells will tend to have relatively constant relationships as a result. The model also implies that these feedback responses may be chained sequentially through a ripple of influences and homeostatic adjustments such that the outermost shell may influence the innermost core, and vice versa, even for the most buffered cases. This means that a changed situation in the outer envelope may have the effect of altering some aspect of the physiological or morphological layer to such an extent that the physiological/ morphological layer forces the shift in a feedback relationship with the genome.

The most obvious evidence for these types of interactive feedback loops comes from the field of psychophysiology where training mechanisms which influence behaviour have been rigorously demonstrated to influence physiology. These phenomena have been demonstrated for laboratory animals (Basmajian, 1963a, 1967; Wells, 1970) as well as humans (Basmajian, 1963b; Headrick, Feather, and Wells, 1970; Miller, 1969; Steptoe, Mathews, and Johnston, 1974). In these and other experiments it was shown to be possible to extend conscious, cognitive control over the physiological processes of the autonomic nervous system at all levels between the single neurone, and complex activities such as heart rate control and regional temperature variation.

Similarly, a degree of morphological change may occur under some conditions as has been shown by Abitbol's examination of ontogenetic change in the human pelvis due to bipedalism (Abitbol, 1988). Long-term morphological change has also been demonstrated for the rhesus macaques of Cayo Santiago, where a secular trend to increased long-bone length and several other size and shape factors has been observed over a number of years (DeRousseau and Reichs, 1987). This aspect of ontogenetic change as a response to external conditions will be examined again below.

Natural Selection and Behaviour

As Mayr has often pointed out (Mayr 1960, 1963, 1970) an organism's behaviour is the first point at which the environmental conditions can act as selective pressures on an organism's genome. In the evolutionary scenario, an organism is constantly being subjected to a multitude of environmental pressures, some of these are intercepted by the behavioural envelope, while others can only be intercepted at physiological or morphological levels of the envelope. That evolution may be driven primarily by the interaction of behaviour and ecological circumstances has been strongly argued by Geist (1972) in an article in which he concludes that the majority of the defining mammalian characteristics of the therapsid (mammal-like) reptiles can be attributed to this process. If the selective pressure is of an acute pattern, offering the organism an immediate or short-term option to survive (and thus maintain the potential for reproduction) or die, it is the levels of cognition, self-awareness or learned tradition which provide the first response. In this sense, being self-aware, cognitive, and having a set of learned responses may be the critical difference between escaping from a selection mechanism such as a predator or becoming the predator's dinner. Conversely, a self-aware, cognitive, traditionally behaving predator also must respond to its own suite of selective pressures, and perform effectively at taking prey. These behavioural responses, on either side, may or may not be supported by physiological variations on a continuing physiological accommodation basis or as an immediate, short-term chemical shift, but the genetic level would not be directly involved. However, it must be recognized that if the organism is selected out of the gene pool there may be some effect on the gene frequencies in the pool, but the effect is relatively greater for a small population than for a large one. If selective pressures are of a chronic, or continuous form, the responses of the organisms may be more generally involved at the behavioural and physiological levels. If the behavioural and physiological layers, acting in a feedback relationship, are able to strike new balances with the changed environmental conditions, for each individual organism, then the population can continue to exist in the same area over a lengthy time period without significant changes in the gene pool, at least with respect to the particular suite of selective pressures, although some change in morphology may occur if that is associated with the appropriate physiological response.

Behaviour and physiology of the individual in the model is thus seen as a complex, interactive, cybernetic system which holds the environment away from direct effect on the genome, yet they respond to and coordinate interactively with the conditions to maximize the possibility of survival for the individual organism. In order to provide an example of the applicability of this model, I will draw upon my own data from a study of behavioural thermoregulation among the Japanese macaques *(Macaca fuscata)* of the Arashiyama A troop. Behavioural thermoregulation involves the use of behavioural activities to aid in the physiological processes of temperature regulation within an organism's body. In the study presented below, the activities of posture change and orientation towards or away from environmental climatic factors was the centre of interest. To prepare for that example it will be necessary to make a diversion into the realms of thermodynamics and ecology.

An Introduction to Thermodynamic Principles

Every animal is a heat engine which takes in both oxygen and oxidizable compounds, performs the oxidation process combining them, and uses a portion of the released energy for activity and body construction purposes. Some of this energy is needed for maintenance of body temperature in homeothermic animals, and some is thrown away as "waste" heat to the environment. The requirement for input energy is very much higher for a homeothermic animal which maintains a relatively constant temperature, than it is for a poikilothermic animal which may allow its internal temperature to fluctuate with the environment. Homeotherms, due to their maintenance of a constant temperature can be consistently active at a constant rate over a broad range of thermal conditions, but there is a substantial cost for this in terms of much higher input requirements, higher rates of oxygen consumption and the need for morphological and physiological mechanisms to aid thermal control. In an overall sense, homeothermy may have some benefits over poikilothermy, but the animal must become more complex, and fragile, in order to enjoy those benefits. For every homeotherm, there is a minimum level of energy utilization at which it is able to maintain a "normal" body temperature. All activity requires additional energy above this Resting Metabolic Rate (RMR), and thermoregulation in conditions outside of a narrow range in which RMR and heat loss are in balance, requires yet more energy.

For every animal there is an environmental "Thermo Neutral Zone" (TNZ) (see Figure 3), outside of which energy metabolism must be increased in order to maintain internal equilibrium. The increase of metabolic rate as temperatures decrease is usually a straight regression line, as illustrated in Figure 3.

Figure 3: Curve of Resting Metabolic Rate for Mammals

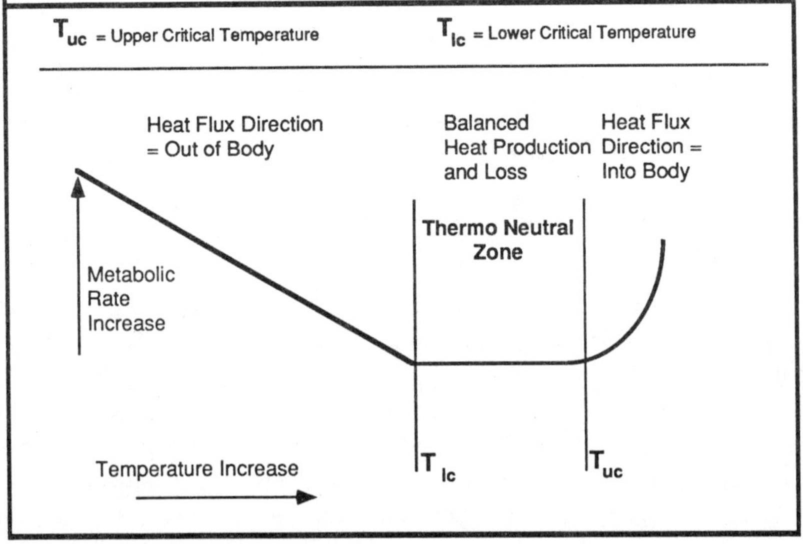

The regression is very dependent upon relative body size and the insulative capacities of the animal's outer envelope, with lower regression coefficients reflecting higher insulative values. This straightforward regression can be substantially modified by the postural behaviour of the animal (Muller and Jaksche, 1980; Paterson, 1980, 1982, 1986). As temperatures rise above the thermo neutral zone, active measures, postural changes, panting or surface perspiration, must be undertaken by the animal in order to reduce the inflow of heat energy from the environment, and to pump internally generated heat against the thermal gradient (Hill, 1975; Swan, 1974; Paterson, 1982).

This avoidance of overheating is the most difficult aspect of thermoregulation and is performed via the mechanisms of respiratory panting or by surface perspiration for evaporative cooling. These processes are energetically expensive, and result in the generation of more heat within the body. This process is a rapid positive feedback cycle which can proceed quickly to thermal overload and failure of the organic structure (Kleiber, 1961; Paterson, 1982).

The Roles of Size, Shape, and Colour in Thermoregulation

Within the processes of thermoregulation, three factors, those of overall size, physical shape, and surface colour play roles in controlling and aiding thermal balance.

Kleiber's Rule states that metabolism (M) is related to mass (W) in the equation:

$$M = b\,W.75 \quad [1]$$

where b is a species-specific constant and W .75 indicates that the mass is taken to the three-fourths power. It is, however, more commonly expressed in Huxley's (1932) logarithmic transformation form of [log y = a (log x) + log b] as:

$$\log M = .75(\log W) + \log b \quad [2]$$

This formulation leads to the conclusion that larger animals are more effective in their thermoregulation than are smaller forms (Reisenfeld, 1981; Schmidt-Nielsen, 1970, 1975; McNab, 1970; Hill, 1975). While an animal cannot violate the physical principles involved in the operation of its energy budget (Campbell, 1977), it is possible to change the internal allocation of energy resources as the animal becomes larger in size. A larger animal is able to reduce the proportion of its energy which is allocated to thermal maintenance since it is able to store larger amounts of heat energy in its body tissues. The larger animal also has a relatively smaller heat-exchange surface since as a consequence of the allometric enlargement, body mass grows faster than surface area. Furthermore the larger animal is able to increase functions such as stride length and rate of travel (Alexander, 1985) which require absolutely larger energy output, but not as much as would be expected from the size increase. That is, the absolute energy requirements for one animal which is twice the size of another are not double but only about one and a half times the amount required by the smaller version. This is viewed as a form of increased efficiency whereby the animal reduces its maintenance costs and increases its capacity for other activities at the same time that it lowers its metabolic rate. It is vitally important not to confuse the issue of **absolute** versus **relative** factors in this. It is certainly true that larger animals require absolutely larger amounts of energy input, but it is also true that if the energy requirements are considered relative to body mass, then a larger animal requires less energy input per kilogram of body mass than a smaller one. A larger animal is thus relatively more efficient in its energy requirements (Wheatley, 1982). Wheatley's example showed that *Pongo pygmaeus* males and male long-tailed macaques *(Macaca fascicularis)* exhibit food requirements of 2,333 kilocalories per day and 855 kilocalories per day respectively. In spite of carrying body weights of 55 and 5 kilograms each, respectively, the food requirement of an orangutan male is only 2.73 times that of the macaque male. The contrast between a body mass of 11 times versus an energy requirement of only 2.73 times is explicative of the importance of allometric scaling to primate ecology. Wheatley also points out that the scaled relationship extends to relative metabolic rate as well, with the "resting daily metabolism per gram of body mass" (Wheatly, 1982: 357) in *Pongo* being approximately half that for *Macaca,* and consequently, the former 'species' daily energetic requirement per gram is about 25% of the macaque" (Wheatley, 1982: ibid.).

Shape, that is, the physical, geometric morphology of the animal related to its surfacace area, is also a factor in thermoregulation, since thermal transfer is an area-related function. That is, more heat can be channelled through a larger surface area than a smaller one, and as Swan (1974) has shown, the variation in shape from a sphere to a complex heteromorph animal shape may represent a substantial change in surface area. A calculation on the basis of the data on surface areas of Japanese macaques *(M. fuscata)* by Hori et al. (1972) yielded the value of 2.536 times the area of a sphere of the same mass (see Table 1 and Paterson, 1986).

Gould (1971) has also pointed out that the "b" value in the allometric equation [log y = a (log x) + log b], of which Kleiber's Rule is just one example, is a "shape factor" in scaled increases in size. He states:

> ...*b is* a scale factor; it shifts as the arbitrary unit of measurement changes and when the size of objects, but not their range of shapes, is altered. In other words, whenever we modify size and *only* size, *b* and *only b* changes. Thus this measure tells us whether a change in b represents a shift of scale alone. If the actual difference in the size of animals on two regressions equals this measure of difference in size for constant shape, then a change in scale alone has produced the shift in b, for the animals on the two curves are constituted in geometric similarity. (Gould, 1971: 120)

In addition to accounting for the cases of organisms which display the same shapes at different size ranges, this provides an important perspective on the thermoregulatory processes of animals. Since shape can be seen as representing the surface area of the animal, and as will be more extensively examined below, the surface area of an animal can be changed by behaviour, there is a substantial linkage between the thermoregulatory capacities of an animal and its body proportions in relationship to size. The colour of the surface exposed to the environment also influences the rate of thermal exchange, and this topic has been extensively examined by Hamilton (1972). The basic model is a straightforward version of the physics "black body - white body" problem of thermal exchange, in which it can be demonstrated that a black body absorbs and radiates heat very much faster than does a white body. Since the radiative absorptivity of a white body is lower than that of a black body, the former heats up less rapidly, but is also much slower at cooling down, thus retaining heat longer. The black body gains and loses heat rapidly, thus does not retain heat for as long. This has been used as an explanation of surface colouration in organisms in differing radiative regimes. Those environments which display a "patchy shade" and short-cycle radiation regime have been shown to favour the development of dark colouration, while those with long cycles in radiation intensity favour lighter colours (Hamilton, 1972). Hence, most tropical forest and tree savanna animals are dark coloured whereas both open grassland savanna and arctic forms are light coloured. The former set of organisms have shade available in which to rapidly radiate heat away, and are able to gain environmental heat rapidly when needed. The latter set either resist heating environmentally, or reduce their radiative heat loss partially through light colouration.

Bergmann's Rule and The Problem of Surface Area

Surface area for any living animal is difficult to accurately measure, sufficiently so that for extensive evolutionary and allometric purposes, exact measurement is considered to be "a fruitless pursuit" by Schmidt-Nielsen (1984). Yet at the same time he states: "in reality the exposed surface area of a living animal changes radically with its posture, whether it curls up or stretches out, how it holds its ears, and so on" (Schmidt-Nielsen, 1984: 82). The measurement of surface area has been conducted primarily by the crude mechanical methods of applying a covering of non-elastic tape or photographic film to a shaved animal (Hori, Tokura and Tadaki, 1972), by employing a modified mechanical revolution counter referred to as a "surface integrator" (Brody, 1945), or by using an integrating planimeter on a fresh skin (William D. Ross, pers. comm.). The precise measurement of surface area is unlikely in any of these methods, for reasons related to problems in defining what the "real surface" of the animal is. Most of the efforts have been directed at the concept of "skin surface area" and have ignored the "hair-top surface area" which may also be important to thermoregulation.

One of the most frequently considered and argued about aspects of the significance of the surface area of an animal comes under the label of the "surface rule." This concept was independently developed by Sarrus and Rameaux (1837), and by Bergmann (1847) who was the more successful at popularizing the idea. This surface rule suggests that a straightforward geometric relationship, that of the surface area to volume relationship known as the "square-cube law," applies as a controlling factor to various physiological processes. The concept of the surface rule with its 2/3 relationship governing size-scaled metabolic processes has gone by the wayside in favour of the 3/4 or mass-scaled relationship known as Kleiber's rule (see above) instead (Schmidt-Neilsen, 1984; Peters, 1983; Kleiber, 1961).

One of the lesser items of Bergmann's work, which has yielded the eco-geographic rule known explicitly as Bergmann's rule, remains a contentious issue in modern ecology. Mayr's restatement of the rule is that "races of warm blooded vertebrates from cooler climates tend to be larger than races of the same species from warmer climates" (Mayr, 1956: 823). Geist in a 1987 paper denies both empirical and theoretical bases for Bergmann's rule. Despite his proofs, he has merely muddied the waters with a significant number of errors in support of an *a priori* conclusion. Geist does provide a summary of the argument about the validity of the rule, but seems to rely heavily and uncritically upon McNab's paper (1971) which makes use of overall length of animals rather than mass or surface area in making his assertion that the rule is invalid. Geist confounds his data set with unrecognized multiple causality, a lack of critical analysis of biological relevance, and misinterprets a physiological relationship which leads him to an argument by ridicule (a detailed refutation of Geist's argument and his rejoinder are forthcoming in the Canadian Journal of Zoology, Paterson, in press). I conclude that Geist's attempt to prove Bergmann's rule invalid has failed (as have the criticisms of Scholander 1955, 1956; Irving 1957, 1972; and Reinig, 1939), but he has pointed to a problem of ambiguity in its expression. Since ecological circumstances and evolution are multicausal complex phenomena, I would propose that Bergmann's rule be restated along the lines of the following:

> 'All other factors remaining equal, within a species, larger massed forms will display a survival advantage leading to increased reproductive fitness over smaller massed forms of the same shape factor in cooler climatic regimes, and vice versa in warmer climatic regimes.'

Body size and shape are complex end products and, like behaviour, are multi-causal in origin. Bergmann's Rule is only one of the set of factors, and can be masked, or overridden by others in differing ecological or climatic regimes.

Another of the eco-geographic rules which should be mentioned is Allen's rule (Allen, 1877) which suggests that the extremities and projections of an animal's body should be smaller and thicker in cooler climates, and larger and thinner in hotter ones. This rule is tied very closely to Bergmann's rule, but has generally been ignored both by those who favour the surface rule, and those who wish to throw it out. (At least I am unable to find any recent consideration of it.) As should be obvious, it is a straightforward recognition that the extremities and projecting structures result in increases in the surface area of the animal, large increases being advantagous under warmer temperatures, while smaller surface areas are of advantage under cold conditions. Animals with long, slim, or thin structural projections, such as the long tail of many primates, are thus at risk under cold conditions. Under conditions of extreme cold, animals may suffer tissue death due to freezing of extremities, unless there is some physiological or behavioural compensatory mechanism. At the South Texas Primate Observatory this has been accidently demonstrated by freezing and resultant loss of tails in the group of vervets (*Cercopithecus aethiops*) during the occasional periods of below 0 C. temperatures. Under these freezing conditions, the ability to maintain acceptable thermal conditions within the vervet's tail is diminished, tissue death occurs, and the tail is lost. Thereafter the vervets have no problems with maintaining body temperatures under minus 0 C. conditions (L. Griffin-O'Neill, pers. comm.). The tail of the vervet monkey forms an effective thermal radiator of large surface area under high temperatures, but at the freezing point the surface-to-volume ratio of the structure is too high to maintain it within a viable range of internal temperature. Similar problems have befallen baboons kept in southern Ontario (Burton, pers. comm.).

The Thermodynamic Situation When Behaviour Is Concerned

Behavioural activity is also a mechanism through which an animal can effectively change its shape, and thus its surface area, hence controlling the major avenues of heat flux between the environment and its body. As has been pointed out (Paterson, 1986; Swan, 1974), the study of these shape changes is a complex problem in descriptive variable and quantitative surface geometry, one which is still in its infancy. Even the mathematical models for it remain based upon straightforward Euclidean geometry (Swan, 1974). For practical purposes it is necessary to construct a set of arbitrary, limited postural patterns which approximate to geometric forms such as spheres, cylinders, "doolittles" - hemispheres on the ends of a short cylinder - prolate and oblate spheroids, ellipsoids, or extremely irregular shapes which can be called heteromorphs (Swan, 1974; Paterson, 1986). Each of these represents a change in relative surface area such that the relationships can be presented as Table 1.

Table 1. Surface to weight or volume relationships of specific shapes. (Assumption of Specific Gravity = 1.0)

Shape	Ratio of surface to the surface area of a sphere of the same weight ("s" or Meeh Constant)	Surface area (S) in dm^2 when W = 1Kg in the Meeh equation $S = s(4.85)W^{.67}$
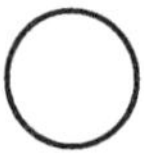 Sphere	1.00	4.85
 Oblate Spheroid	1.09	5.29
 Prolate Spheroid	1.10	5.34
 Doolittle	1.12	5.43
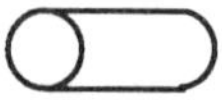 Cylinder	1.18	5.72
 Ellipsoid (face and cross-section)	1.30	6.30
 Heteromorph	1.70	8.30
Macaca fuscata (Hori, et al,1972)	2.536	12.30

Since most animals have proportional surface-area values (right column) in the 9.0 to 13.0 square decimetre range (although Schmidt-Neilsen (1984) has shown that bats, *Chiroptera*, at 57-58 dm^2 are at the extreme outside range,), it is obvious that behaviour which changes the surface configuration presented to the environment towards these geometric forms will yield a reduction in surface area and reduced heat flux (transference of thermal energy, either in or out).

Posture: Flexion Creases and Surface Area

The postures which animals may assume are influenced by a multitude of variable factors, but irrespective of their complex origins, they result in changes in the rate of energy interchange with the environment. The four channels, radiation, conduction, convection, and evaporation have a common feature, they are all described by the rules of Newton,[2] Stefan-Boltzmann,[3] and Fournier,[4] and are surface area dependent. The postural changes of an animal modify the effective surface area, the "physiological surface" (Brody, 1945), through approximation of the various flexion creases of its body. These are the axillary crease, the elbow, knee, hip, etc., one, two or three for each joint of the body. One could construct a very detailed action matrix dealing with each flexion crease and the degree to which it is open or closed, and that may be useful for a micro level study of thermal interchange, but is likely to be trivial in terms of large-scale results. The observational categories employed in a study of *Alouatta palliata mexicana* (Paterson, 1982, 1986) have been modified for use with *Macaca fuscata* at the South Texas Primate Observatory, and the current set of postures is illustrated in Figure 6.

Figure 6: Postures Employed by Japanese Macaques

An alternative set of postural categories has been employed in the study of baboon *(Papio cynocephalus)* thermoregulation by Stelzner (1987; Stelzner and Hausfater, 1986).

The Role of Position and Orientation Within the Environment

It has also been shown that animals can alter their thermal exchange pattern with the environment through locational and orientation factors within the environs. Thus they can be shown to have preferences for particular microhabitats, and consistent patterns of orientation towards environmental factors. Sunning behaviour and shade seeking have been studied in *Macaca mulatta* (Dahl and Smith, 1982; Dahl et al, 1986) and wind speed considered to be an important factor in their results. Sheltering from the wind and orientation of a wind-on-dorsum posture have been noted as important aspects of the thermoregulation of the koala *(Phascolarctos cinereus)* (Degabriele, 1980; Degabriele and Dawson, 1979). Stelzner has found opportunistic utilization of shade and differentiated progress rates in differing shade regimes in a field study of baboon thermoregulatory behaviour (Stelzner, 1988). As I noted in 1986:

> Microhabitat preferences allow the utilization of those variable factors of the environment (solar radiation, environments re-radiation patterns, air flow rates, relative humidity, etc.) which can increase or decrease an animal's rate of thermal exchange. Microhabitat factors for terrestrial and arboreal environments are obviously very different (see Paterson, 1982), and in consequence the two major habitat conditions of primates involve substantially different thermoregulatory microhabitat preferences. The location and orientation chosen by a primate can serve to help in the maintenance of normal body temperature through manipulation of the microclimate. Preferential factors can be observed easily in the relatively inactive arboreal primates such as *Alouatta* and *Colobus* where often apparently resting animals will suddenly change their locations and resume the same postures. Such changes may indicate a taxis toward a microhabitat in which they may maintain metabolic thermoneutrality more easily. (Paterson, 1986: 236).

A Case Study: Arashiyama Male Macaques

At the time of my visit in 1980 to the Arashiyama A troop at their original location near Laredo, Texas, I gained the impression, and Claud Bramblett (pers. comm.) confirmed that the older males, born and partly raised in Japan, were of a distinctly different body conformation than the younger males who had grown to adulthood in Texas. These young males had bodies approximately 10% longer in trunk length, and presented a picture of a more linear or greyhound-like form compared to the burly bear-like configuration of their elders. An initial hypothesis might have been that they were being fed better, and hence were reaching their genetically determined maximum sizes. This may be partially true, yet the annual data collection showed no significant differences in weight (Bramblett, pers. comm.), thus suggesting a multi-causal model should be employed. My initial thoughts were of the "dolichocephalic-brachycephalic" contrasts and size changes seen in European human families after their migration to North America (Boas, 1911 noted by Harrison, et al., 1988), but my immediate response was to consider these conformation changes as a potential test example for the functioning of eco-geographic rules involving surface area changes such as Bergmann's and Allen's rules.

The thermal conditions, humidity, precipitation, wind, and cloud cover conditions in south Texas are substantially different from Japan, and if Bergmann's-Allen's rules had any measurable effect, I hypothesized that there should be observable differences in the behaviour patterns and responses to environmental factors, and especially in the postures and orientations employed by 'long,' younger adult males and 'short,' older Japan-born males. Through five weeks of July and early August, 1982, part of what is on record as one of the hotter summers of the decade, I conducted field research with a selected set of these adult males at the new Arashiyama location near Dilley, Texas.

Methods and Materials of Study

The study was conducted using the Focal Time Sampling procedure of Baulu and Redmond (1978) in order to gain a maximal amount of statistically usable data in the limited time available. Five-short and five long-bodied males were selected as the focal animals of the study. They carried the AWI tattoo numbers 1, 28, 64, 128, and 134, and numbers 2, 55, 74, 129, and 143 respectively (see Table 2).

Table 2: Sample Males in Postural Study

Tattoo Number	Common Name	Family	Birth Year
	Short Bodied Males		
001	Fatboy	Rheus	71
028	Suma	Suma	64
064	Moe	Momo	64
128	Meme65 (Hulk)	Meme	65
134	Ran68	Ran	68
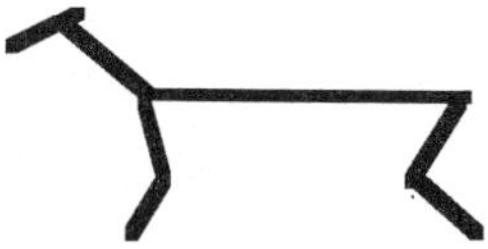	Long Bodied Males		
002	Dimbulb 2	Pelka	70
055	Wania One Eye	Wania	70
074	Fang	Rotte	70
129	RockyTwo	Betta	71
143	Dimbulb 1	Pelka	68

All were mature males 10 years of age and older during the study. At the time of writing (1988-89), four of the ten males have died, mostly from *Coccidiodomycosis* (Valley Fever), and Fatboy, who has recovered from the same disease, has changed body configuration towards the long-bodied form through the apparently permanent loss of 25% of his body mass. The study collected a focal time sample at 2 minute intervals throughout daylight hours for a total sample of 13 hours for each male. The sample records consisted of the subject's:

Behavioural Activity, scored on a limited set of state behaviours (Altmann, 1972), recorded and employed in the analysis as nominal values;

Posture assumed, ordered into an increasing surface area scale (Paterson 1982, 1986, and Figure 6), constituting an ordinal scale;

Amount and Orientation of body exposed to the Sun, exposure measured in terms of fifths of surface area and anatomical surface which is oriented to sun, a combination of ordinal and nominal values;

Amount and Orientation of body exposed to any Wind flow, exposure measured in terms of fifths of surface area and anatomical surface which is oriented to wind, again a combination of ordinal and nominal values;

Altitude above ground to nearest tenth of a metre, thus recorded as an ordinal scale;

Substratum type supporting the subject, vegetation, structural materials, variations of ground type, etc,. roughly ordered in terms of increasing thermal coefficients, thus again constituting an ordinal scale, and

the shade **Temperature** at human chest height (approximately 1.3 metres), later grouped into 5 C. **Temperature blocks,** another ordinal scale

The postures are ranked on an ordinal scale from most spherical, the ball, displaying the least surface area, to most surface area exposed to the environment, the open sit and quadrupedal stand. This ordinal scale is used in the charts presented as Figures 4 and 5, as the x-axis.

Figure 4: Graph of Postures and Temperature Blocks for Short Bodies

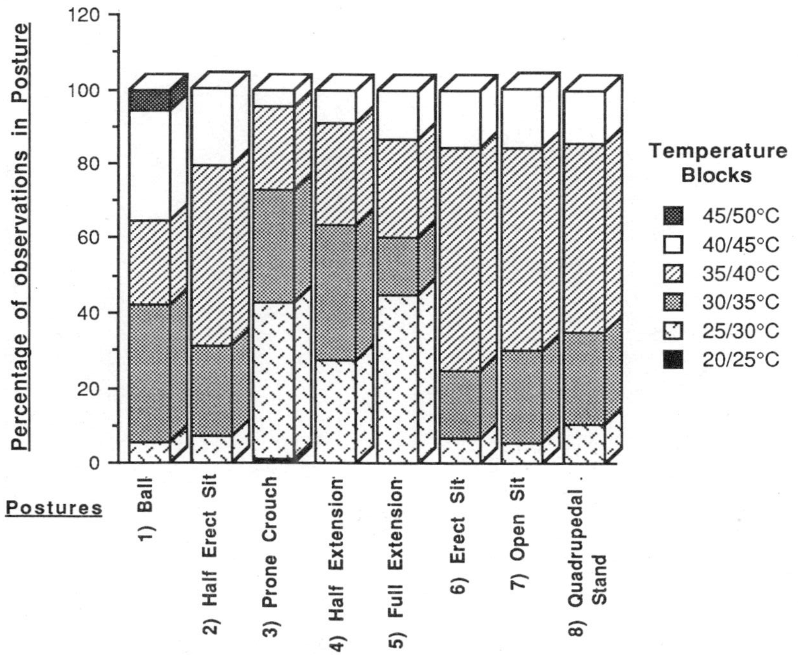

Figure 5: Graph of Postures and Temperature Blocks for Long Bodies

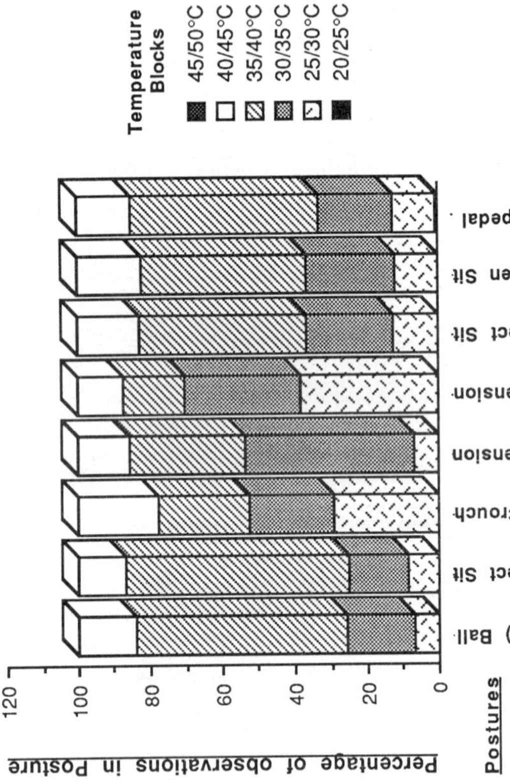

Each record also included time-control markers, a measure of the amount of sky clouded, estimated in tenths, precipitation measurement - of which there was none throughout the period - and wind velocity data, as well as the subject's identity. The data were collected using a specialized program on a Radio Shack Pocket Computer Model 1, subsequently stored on tape and later transfered to a microcomputer system (see Paterson 1988 for a description). Data analysis was conducted on a Macintosh Plus with the aid of the Statview 512+ statistical package.

Hypotheses:

The "working hypotheses," constructed in advance for this study were as follows:

[1] Postures with lowest surface-area-to-volume ratios should increase in both frequency and duration of use under conditions of low temperature, and again under the highest temperatures.

[2] Posture utilization frequency should reflect an inverted version of the RMR base curve (see Figure 3) to some extent, with little consistency in the Thermo Neutral Zone, but with increasing use of maximum area/volume postures as temperatures rise, until a shift to minimum area/volume postures as the point of cooling failure is approached. Low values, but high significance for correlations and regressions should be expected since postures are also influenced by behaviour (both the subject's and nearby individuals'), internal heat production, humidity, social interactions, locomotion, lag times and refractory periods for cooling and digestive statuses, as well as a set of unknown factors.

[3] In order to obtain useable regression and correlation values of biological significance, it would be necessary to partition the data in a number of different ways during analysis. (It seemed likely that resting activity or the set of relatively inactive behaviours would likely be the situation which produced the most significant results, but it was also

considered to be necessary, for reasons related to the physiological curve (see Figure 3), to keep the data from higher temperatures separate from that for lower temperatures.

[4] Sun exposure should be positively correlated with low temperatures, while it should be negatively correlated with high temperatures.

[5] Wind exposure should be negatively correlated with low temperatures, while it should be positively correlated with high temperatures.

[6] Substrata with higher thermal conductivity should be selected when heat loss is required. This should occur at high temperatures, and after activities which involve the generation of more internal heat.

[7] Altitude above ground should be selected as a cooler environment, but should have a negative effect on posture.

[8] Wetting of the body, or swimming should be mechanisms used to rapidly cool off, thus they should be utilized more at high temperature than at low temperature (at least under the Texas conditions).

Only the working hypotheses 1 through 7 will be considered in this paper. The initial inclusion of hypothesis 8 was made with the expectation that swimming would be recorded for the subject males, since juveniles of both sexes and adult females were noted to swim at Laredo in 1980. However, in this study, fully adult males were never observed to go swimming. Not even on those occasions when air temperatures rose above 45 C. and large numbers of infants, juveniles, and adult females were in the water. The nearest to adult male swimming was one case in which Moe (#64) waded into the pool until his hands and feet were covered and remained there for 90 seconds. In *ad lib.* observations between samples, however, it was observed that Wania One Eye (#55), a low-ranking and chronically ill individual, would regularly dunk his lower body into a water tank during the thermal peak of mid-afternoons. Postures are the result of a very complex interplay of a number of internal and external factors, but one of the more significant of these is behavioural activity. It may seem to be intuitively obvious that this relationship between posture and activity exists, but in order to be certain, I tested the hypothesis in the data and found a contingency coefficient of .708 between them, with a Cramer's V of .354, and Chi-Square significance at the $p < .0001$ level over the entire data set.

Results

Since the first objective of the study was to determine relationships between body form, postures, and temperatures, contingency tables were calculated for posture with temperature in 5 C blocks for each body type (Table 3).

Table 3: Contingency Table

Body type	Contingency Coefficient	Cramer's V	χ^2 p
Posture and Temperature Blocks (5°C units over 20 to 50 °C.)			
Short	.425	.210	<.0001
Long	.282	.147	<.0001
Posture and Sun Exposure			
Short	.488	.250	<.0001
Long	.425	.210	<.0001
Posture and Wind Exposure			
Short	.441	.186	<.0001
Long	.470	.201	<.0001
Posture and Activity			
Short	.709	.380	<.0001
Long	.713	.384	<.0001
Posture and Substratum Type			
Short	.417	.174	<.0001
Long	.528	.235	<.0001

Similarly, contingency coefficients were processed for the relationship of posture to sun exposure, wind exposure, activity, and substratum type. Note that the contingency coefficient is usually higher for short bodies.

In order to evaluate the differences between the Long and Short bodied individuals, the results of a contingency analysis have been graphed to show the relationships between the percentages of each posture (except the quadrupedal stand) adopted at each temperature block. An inspection of the charts will indicate that there are some obvious differences between the long- and short-bodied individuals in their postural use at various temperatures.

The next stage of the analysis moved to the use of regressions in conjunction with partitioning the data at several stages. At this point it is worthwhile to take careful note of Schmidt-Nielsen's cautions that "*...the proper statistics are no more than a description of the numbers at hand...*" and that *"Highly significant statistics do not signify equally high biological significance"* (Schmidt-Nielsen, 1984: 22). It is the responsibility of the researcher to ascertain the biological relevance of a particular statistic, and to decide the levels of significant variation which will be accepted or rejected. Several forms of regression analysis were tested on the data set, but only the results of the Stepwise Regressions are to be presented here. The purpose of a regression analysis is the generation of a formula which can predict to some (hopefully reliable) extent, the value of the dependent variable, if the values of the independent variables are known. In this case, the purpose is to find the contributing factors among the independent variables which will yield the dependent, that is, the Posture assumed. The equations which are derived from these analyses have the format of:

$$Y = B_o + B_1 (Var_1) + B_2 (Var_2) + B_3 (Var_3) ..... + B_n (Var_n) + \varepsilon \quad [3]$$

or to use the format in which the equations and data will appear:

$$Posture = INTERCEPT + Beta_1(Variable_1)....+Beta_n (Variable_n). \quad [4]$$

The "Beta" component of the equation is the value reported in the tables which follow; the ε or error will not be reported. The calculation which is utilized here is the "Stepwise" Regression in which each variable is entered only once, in order of the calculated significance of its contribution towards the final value of the dependent variable (this calculation is invisibly performed by the analysis program). The INTERCEPT is the value for the posture when the values of the independent variables are set to zero, it represents the starting value for posture, followed by the additive influences of each variable. This calculation yields different equations at each stage, variables being added or dropped depending upon the recalculation of the matrix after each step. Stepwise Regressions can be run with different partitions of the data and will yield different equations. The advantage for these particular analyses lies in the validation of each variable's effect in the equation, something which is not true of Multiple Regression techniques which enter all variables at once, and then provide "partial F" values for the researcher to make a decision as to the significance of the impact. One of the items in each of the following tables which should be noted is the R^2 value which represents the proportion of the variance in the dependent variable which is accounted for by the equation. This can be converted to percentage by shifting the decimal point two places to the right. Since posture has such a complex causality, only a small portion of the total variance is accounted for in any of these analyses.

In many of the equations reported below, the beta values are negative, a phenomenon which might be unexpected, but which is actually a reasonable result. If one considers what the beta value does to the equation, it will be more comprehensible. The beta value is essentially a "weighting factor" which reflects the impact of the variable on the calculated slope of the regression line, and thus may have either an increasing effect, with a positive value, or a decreasing influence when it has a negative value. Thus, positive beta values will tend to increase the slope of the regression, and negative beta values will decrease the slope of the line.

The first reported calculation (Table 4) partitioned the data by body form - the short and the long - and was run for the complete observed temperature range, with 5 physical variables being entered. The variables used were: Temperature Block, Sun Exposure, Wind Exposure, Substratum

Table 4: First Stepwise Regression
Over Full Temperature Range (20-50°C), All activities included, Variables = Temperature Block, Sun Exposure, Wind Exposure, Substratum Type, Altitude.

Short Bodies	Step.1	Step.2	Step.3	Step.4	Step.5
R =	.192	.232	.263	.268	.285
R^2 =	.037	.054	.069	.077	.081
INTERCEPT	4.405	4.037	4.12	3.717	4.38
Temperature Block					–.158
Sun Exposure	.191	.151	.145	.132	.118
Wind Exposure		.104	.117	.114	.116
Substratum				.073	.066
Altitude			–.402	–.464	–.466

Long Bodies	Step.1	Step.2	Step.3	Step.4	Step.5
R =	.295	.306	.31		
R^2 =	.087	.094	.096		
INTERCEPT	3.923	4.019	3.88		
Temperature Block					
Sun Exposure	.296	.297	.286		
Wind Exposure			.037		
Substratum					
Altitude		–.173	–.19		

Equations Yielded at the final step:

Short bodied males: (five steps)

Posture = 4.38 + .118(Sun Exp) + .116(Wind Exp) -.402(Altitude) + .073(Substratum) -.158(Temp.Block)

Long bodied males: (three steps)

Posture = 3.88 + .286(Sun Exp) -.19(Altitude) + .037(Wind Exp) type, and Altitude. The set was not partitioned for activity, hence all activity patterns are included, but Activity is not entered as a variable. Since the minimum value of the F-ratio used to enter all variables into these regressions was set at 4.000, all of the reported coefficients are significant to at least the .005 level, suggesting that they have important influences on the postures adopted by the subjects.

This indicates that for long-bodied males 9.6% of the variance in posture use can be accounted for by three variables, two of which, Sun Exposure and Wind Exposure, have a positive impact and one, Altitude, has a negative influence. For the short-bodied males, 8.1% of the variance is accounted for by 5 variables, two of which, Temperature Block and Altitude have relatively large negative influences, while three, Sun Exposure, Wind Exposure and Substratum are positive. The equations yielded are:
for short bodied males:

$$\text{Posture} = 4.38 + .118(\text{Sun Exp}) + .116(\text{Wind Exp}) - .402(\text{Altitude}) + .073(\text{Substratum}) - .158(\text{Temp. Block}) \quad [5]$$

for long-bodied males:

$$\text{Posture} = 3.88 + .286(\text{Sun Exp}) - .19(\text{Altitude}) + .037(\text{Wind Exp}) \quad [6]$$

If, in the interests of a direct comparison, we look at the first 3 variables entered in the short-body equation, they are the same as for long-bodied individuals, with 6.9% of variance accounted for, and lower impacts for Sun and Wind Exposure, but much stronger influence of Altitude than in the long-bodied case. However, the conclusion must be that short-bodied males are influenced by more factors in their assumption of postures than are the long bodied individuals.

The next stage of the analysis partitioned the data set further. For Table 5, the Short and Long partitions were combined with three temperature partitions. These are firstly, the range of 20 to 30 C. which roughly corresponds to the Thermo Neutral Zone (hereafter noted as "TNZ"). Elizondo, (1977) determined the Thermoneutral range for *Macaca mulatta* (the rhesus macaque) from central India to run approximately from

Table 5: **Second Stepwise Regression**: Partitioned by Body Type and Temperature Ranges of Thermoneutral Zone (TNZ) (20-30 ° C), Above TNZ (30-50 ° C), and Above Body Temperature (T_b) (35-50 °C). Variables = Temperature block, Sun Exposure, Wind Exposure, Substratum Type, Altitude, Activity.

Short Bodies	Step1			Step2			Step3			Step4			Step5		
	TNZ	>TNZ	>T_b	TNZ	>TNZ	>T_b	TNZ	>TNZ	>T_b	TNZ	>TNZ	>T_b	TNZ	>TNZ	>T_b
R =	.251	.217	.245	.305	.266	.292		.287	.306		.299	.321		.305	
R^2 =	.063	.047	.06	.093	.071	.085		.082	.094		.09	.103		.093	
INTERCEPT	5.787	4.286	4.214	6.521	3.843	3.8		3.916	3.856		3.526	3.44		3.428	
Temp. Block															
Sun Exposure		.221	.24	.183	.171	.183		.165	.168		.152	.15		.152	
Wind Exposure					.123	.125		.135	.141		.13	.133		.124	
Substratum											.074	.084		.072	
Altitude	-.877			-.8				-.34	-.333		-.413	-.426		-.391	
Activity														.056	
Long Bodies															
R =	.236	.308	.319	.294	.341	.35			.358			.363			
R^2 =	.056	.095	.102	.086	.116	.123			.128			.132			
INTERCEPT	4.72	3.851	3.728	4.978	3.467	3.37			1.738			1.42			
Temp. Block									.383			.393			
Sun Exposure		.316	.324		.314	.312			.314			.309			
Wind Exposure															
Substratum												.045			
Altitude				-.284											
Activity	.291			.288	.184	.194			.195			.19			

Equations Yielded at the Final Step:

For Short bodies:

TNZ Posture = 6.521 -.8(Altitude) +.183(Sun Exp)
Above TNZ Posture = 3.428 +.152(Sun Exp) +.124(Wind Exp) -.391(Altitude) +.072((Substratum) +.056(Activity)
Above T_b Posture = 3.44 +.15(Sun Exp) +.13(Wind Exp) -.426(Altitude) +.084(Substratum)

For Long bodies:

TNZ Posture = 4.978 +.291(Activity) -.284(Altitude)
Above TNZ Posture = 3.467 +.314(Sun Exp) +.184(Activity)
Above T_b Posture = 1.42 +.309(Sun Exp) +.19(Activity) +.393(Temp. Block) +.045(Substratum)

25 through 30 C, but due to the greater insulation factor, the denser, longer hair of *M. fuscata* (the Japanese macaque), the Thermo Neutral Zone undoubtedly extends to lower temperatures for this species. The second partition is the temperature range "above the TNZ" (noted as "Above TNZ," or ">TNZ") from 30 to 50 C And thirdly, the partition representing ambient temperatures "above body Temperature" (noted as "Above Tb," or ">Tb") which is taken as the range from 35 to 50 C, due to an insufficiency of variance in one partition when only the 40 - 50 C. blocks were used. These partitions were run with 6 variables - the same set as in Table 4 - with the addition of Activity as an entered variable.

For long-bodied subjects, there are substantial changes in the percentage of the variance in Posture which is accounted for in the equation. The results indicate 8.6% covered in the TNZ set, 11.6% in the Above TNZ set, and 13.2% in the Above body temperature set. For the short-bodied individuals, the percentage of the variance in Posture accounted for was a much more consistent 9.3% in the TNZ set, 9.0% in the Above TNZ set, and 10.3% in the Above Tb set. Yet for each condition, there were substantial changes in the number of steps, the variable influences, and the directions of influence. The final equations were arrived at after differing numbers of steps in the Long and Short partitions. For the Short bodies, the equations involve 2 steps in the TNZ set, 5 steps in the Above TNZ, and 4 steps in the Above Tb set. For the Long bodies, they ended at 2, 2, and 4 steps for the same partitions. Once more, the short-bodied individuals have a larger number of influences on their postures. The equations produced are shown at the bottom of Table 5.

As can be seen, in the TNZ set, the intercept points have relatively high values which correspond with a "prone-extension" and an "half extension" posture respectively for the short and long bodies. In the Above TNZ set there is a very substantial transformation of influences. Intercepts for both body types fall into the "prone crouch" range. It is apparent that again long bodies have a short list of influences and short bodies have a more extensive listing. In the Above Tb set, there is a substantial shift of intercept for the long bodies, down to the "ball" posture, and the appearance of Temperature Block as a significant factor, the only circumstances in which it does so.

In the third and fourth stepwise regressions (Tables 6 and 7 respectively), the partitioning is by body type (short and long bodies), the same three thermal ranges, but is restricted to the set of inactive behaviours for the third calculation (Table 6),

Table 6: Third Stepwise Regression: Inactive Behaviours: = Resting, Grooming, Receive Groom, Huddling, Sunning, and Resting/Chewing. Partitioned by Body Type and Temperature Ranges of Thermo Neutral Zone (TNZ) (20-30 ° C), Above TNZ (30-50 °C), and Above Body Temperature (T_b) (35-50 °C). Variables = Sun Exposure, Wind Exposure, Substratum Type, Altitude.

Short Bodies	Step1			Step2			Step3		
	TNZ	>TNZ	>T_b	TNZ	>TNZ	>T_b	TNZ	>TNZ	>T_b
R =	.385	.206	.23	.419	.245	.282		.267	.309
R^2 =	.148	.042	.053	.176	.06	.079		.072	.095
INTERCEPT	6.797	3.922	3.889	6.829	3.344	3.223		3.24	3.091
Sun Exposure	-.368			-.342					
Wind Exposure		.136	.149		.121	.124		.133	.142
Substratum					.101	.125		.127	.155
Altitude				-.498				-.283	-.386
Long Bodies									
R =	.164	.243	.269	.217	.265	.321		.272	.327
R^2 =	.027	.059	.072	.047	.07	.103		.074	.107
INTERCEPT	5.277	3.826	3.683	4.656	3.374	2.963		3.329	2.917
Sun Exposure		.249	.282		.221	.231		.225	.226
Wind Exposure				.118					
Substratum					.074	.121		.09	.14
Altitude	-.224			-.276				-.109	-.117

Equations Yielded at the final step:

For Short bodies:

- TNZ — Posture = 6.829 -.342(Sun Exp) -.498(Altitude)
- Above TNZ — Posture = 3.24 +.133(Wind Exp) +.127(Substratum) -.283(Altitude)
- Above T_b — Posture = 3.091 +.142(Wind Exp) +.155(Substratum) -.386(Altitude)

For Long bodies:

- TNZ — Posture = 4.656 +.118(Wind Exp) -.276(Altitude)
- Above TNZ — Posture = 3.329 +.225(Sun Exp) +.09(Substratum) -.109(Altitude)
- Above T_b — Posture = 2.917 +.226(Sun Exp) +.14(Substratum) -.117(Altitude)

and resting alone for the fourth regression (Table 7).

Table 7: Fourth Stepwise Regression: Resting Behaviour Alone: Partitioned by Body Type and Temperature Ranges of: Above TNZ (30-50 °C), and Above Body Temperature (T_b) (35-50 °C). Variables = Temperature Block, Sun Exposure, Wind Exposure, Substratum Type, Altitude.

Short Bodies	Step 1		Step 2		Step 3		Step 4	
	>TNZ	>T_b	>TNZ	>T_b	>TNZ	>T_b	>TNZ	>T_b
R =	.206	.263	.262	.305	.279	.331		.342
R^2 =	.042	.069	.069	.093	.078	.109		.117
INTERCEPT	3.837	3.273	3.169	3.028	3.059	2.892		4.31
Temp. Block								-.32
Sun Exposure								
Wind Exposure	.134		.116	.101	.127	.121		.118
Substratum		.182	.117	.158	.143	.186		.18
Altitude					-.237	-.364		-.342
Long Bodies								
R =	.318	.335	.353	.394	.359	.416		.422
R^2 =	.101	.113	.125	.155	.129	.173		.178
INTERCEPT	3.533	3.409	2.892	2.578	2.833	.391		.302
Temp. Block						.512		.52
Sun Exposure	.293	.319	.253	.263	.255	.27		.265
Wind Exposure								
Substratum			.107	.141	.127	.14		.163
Altitude					-.109			-.118

Equations Yielded at the final step:

For Short bodies:
Above TNZ Posture = 3.059 +.134(Wind Exp) +.143(Substratum) -.237(Altitude)
Above T_b Posture = 4.31 +.18(Substratum)+.118(Wind Exp) -.342(Altitude)-.32(Temp. Block)
For Long bodies:
Above TNZ Posture = 2.833 +.255(Sun Exp) +.127(Substratum) -.109(Altitude)
Above T_b Posture = .302 +.265(Sun Exp) +.163(Substratum) -.118(Altitude) +.52(Temp. Block)

The variables of Activity and Temperature Block (for Table 6) are not in the tables: Activity because it is one of the partitions, and Temperature Block did not enter the regression of either long or short partitions at any step in the third regression. The third regression yielded equations for 2 steps in the TNZ set, and 3 steps in the other two sets for both major partitions. There are, however, substantial differences in the variables contributing to the equations, and the strengths of their influences. The final equations from the third regression are below Table 6.

In examining Table 6 it is possible to note that in all cases except for the short body-high temperature situation, the percentage of the variance accounted for declines slightly from Table 5. Also, in all cases except the long body-Above Tb case, all of the intercept values move slightly lower, indicating lowered surface-area exposure to the environment. Other than these effects, the removal of the variable, Activity, and the restriction to the inactive behaviours partition, seems primarily to weaken the effects of the contributing variables.

In the case of the final regression presented in Table 7, there are again substantial changes in the intercepts, the number of steps in the equations, and shifts in the impacts of the variables. Of importance are the shifts in percentage of variances accounted for, with 7.8% and 12.9% for short and long respectively in the Above TNZ category, and 11.7% and 17.8% in the Above Tb set. The TNZ partition could not be calculated for either short or long bodies because there were too few cases; however, both the Above TNZ and Above Tb situations yielded 3-and 4-step equations for both major partitions. The final equations from the fourth regression are below Table 7.

There is a remarkable similarity of form in the equations with Substratum, Altitude, and Temperature Block occurring in both sets, but the unusual feature is the consistent effect of Wind Exposure on the Short bodies, and Sun Exposure on the Long bodies. In both cases there is no effect on the other partition for each of these variables. The change in intercept level is towards more spherical postures of "prone crouch" and "half erect sit" for most of the steps, but significantly down, well into a "ball" for the long bodies at the third and fourth steps of the Above Tb regression.

Conclusions

In evaluating the working hypotheses, a number of problems arise both from the data set and from the research design in addition to the direct factor of interpretation. In part the "working hypotheses" were not formulated as precisely and specifically as they should have been. As a consequence, four of the hypotheses are acceptable, having failed to be rejected, and three are rejected even though there is evidence of "trends" in the data (see Table 8).

Table 8: Hypothesis Evaluations.

Hypothesis	Evaluation
1) That postures with lowest surface area to volume ratios should increase in both frequency and duration of use under conditions of low temperature, and again under the highest temperatures.	1) Reject: the frequency data for the two body types is not completely comparable, there is an absence of cases for the long body data set in the temperature blocks below 25 ° and above 45° C. The data suggest a tendency to follow the predicted pattern, but it is more evident in the short bodies than in the long individuals.
2) Posture utilization frequency should reflect an inverted version of the RMR base curve (see Figure 3) to some extent, with little consistency in the thermoneutral zone, but with increasing use of maximum area/volume postures as temperatures rise, until a shift to minimum area/volume postures as the point of cooling failure is approached.	2) Fail to Reject: The correlations and regressions do in fact show low values with high significance, confirming that there are a multitude of factors, including unknown influences which are interacting in the animal.
3) That in order to obtain usable regression and correlation values of biological significance, it would be necessary to partition the data in a number of different ways during analysis. (It seemed likely that resting activity or the set of relatively inactive behaviours would likely be the situation which produced the most significant results, but it was also considered to be necessary, for reasons related to the physiological curve (see Figure 3), to keep the data from higher temperatures separate from that for lower temperatures.	3) Fail to Reject: It is necessary to carefully partition the data set to obtain reasonable results, but often the results are unexpected. It is also true that partitions separating the resting and inactive behaviours produce good equations, but not significantly better than a regression in which activities are included as a variable.
4) Sun exposure should be positively correlated with low temperatures, while it should be negatively correlated with high temperatures.	4) Reject: Sun exposure seems to be of major importance only in the long bodied data set, where the influence is fairly consistently present and positive for all data sets above the TNZ in regressions number two, three and four. Sun exposure is much less consistent for the short body data set where the strength of its influence is less than half that for long bodies.
5) Wind exposure should be negatively correlated with low temperatures, while it should be positively correlated with high temperatures.	5) Reject: Wind exposure does display a moderate influence throughout regressions two, three, and four, but only in the short bodied individuals. Wind exposure seems to be irrelevant to the long bodied macaques.
6) Substrata with higher thermal conductivity should be selected when heat loss is required. This should occur at high temperatures, and after activities which involve the generation of more internal heat.	6) Fail to Reject: Substratum is an erratically appearing factor in the equations but does display a tendency to provide a modest positive influence in the higher temperature partitions for both long and short bodies in regressions three and four, the inactive and resting data sets respectively.
7) Altitude above ground should be selected as a cooler environment, but should have a negative effect on posture	7) Fail to Reject: The influence of altitude is present in regression two, having major impacts on short bodies at all temperature ranges, but only appears once in the long bodied set. Its influence is more consistent in regressions three and four, for the higher temperature sets and is substantially more important for the short body equations.

In particular, hypothesis four, concerning the importance of Sun Exposure, while rejected on the overall data set, could be accepted for the case of the long-bodied individuals. It is evident that more attention should have been paid to the hypothesis-construction phase of the research design, an activity which might have been rewarded through the existence of a larger number of explicit hypotheses for testing. In addition, it is evident that the data set must be expanded through the acquisition of records from a larger sample of individuals and from all seasons of the year.

An obvious and important conclusion which can be drawn from the analyses presented here is that environmental factors affect the two body types in different ways, and that the long-bodied males are less obviously influenced by environmental factors. It is also possible to conclude, but not confirm at this time, that short-bodied males have greater problems in regulating their body temperatures under the stresses of the high temperature ranges, and hence are maximizing the utilization of as many environmental variables as possible in order to avoid heat prostration. The long-bodied males do not seem to need these extra mechanisms, though the surprisingly low intercept value, in the range of the "ball" posture found in regression four for the above body temperature (>Tb) partition indicates that postural change may be of major significance to them when temperatures rise above 40 C. With the relatively larger surface area of the long bodies, once the temperature conditions move into the extremely high ranges, that increased area becomes a handicap, and retreat into a spherical posture may be an effective mechanism to decrease heat flux inflow. Since the change in thermal conditions for the Arashiyama A troop has been adjusted upwards from Japanese conditions both in terms of the yearly mean averages and the summer highs, the increased surface area of the long-bodied males reflects adaptation to those new conditions. Even though they may have some problems at the extreme high temperatures, they have managed to escape many of the environmental influences through this plasticity of ontogenetic development. The troop has been moved from a condition, at Arashiyama Mountain, where summers are pleasantly warm, and the winters can be cold and stressful, to southern Texas where the winters are mild to warm, and the summers are very hot and stressful. Under these contrasting conditions, it is appropriate to view the normally short-bodied form of the Japanese macaque as adapted to the colder climate of central Honshu, and the longer leaner form as a plastic adaptation to the hotter environment in Texas. I therefore formally conclude that the long-bodied males reflect an adaptation to the thermal conditions of south Texas, and provide an example of developmental plasticity of physiology and morphology, which can be perceived through the differentiation in patterns of behaviour, particularly the use of postures.

The manner in which the differences between the two body types is manifested can be seen in the ways in which the macaques behave. When conditions are in the Thermoneutral range (20 to 30 Celsius), short-bodied individuals tend to assume postures of the "erect sit" form, and use a little exposure to the sun, while long-bodied individuals adopt "full extension" postures, influenced partly by their behavioural activity. At the other extreme, when ambient temperatures are higher than their body temperatures, short-bodied males adopt "prone crouch" postures which are influenced by sun, wind, substrate, and altitude factors, while long-bodied males take up a "ball" posture which is influenced by temperature, sun exposure, and substrate. Overall, the postures used by short-bodied individuals reflect more involvement with environmental factors in the 20 to 40 C range than is apparent for the long-bodied males, hence it may be concluded that the short-bodied animals are using these factors to maintain a constant internal temperature, while the long-bodied individuals are more consistently in thermal balance.

The topic of postural thermoregulation is of significance as a mechanism for the examination of proximate causality of behaviour, and may prove to have an important relation to the ultimate causality of body shape evolution. However, more investigation, a larger data set, and especially acquisition of internal thermal conditions from a set of subjects under a broader temperature range will be necessary before more extensive conclusions are justified.

Summary

In the course of this chapter, I have presented a model for the proximate causality of behaviour and tried to substantiate the roles of each of its layers. I have also shown how physiology and morphology can be correlated with behaviour, especially within the realm of thermal biology. The examples from the field study of behavioural thermoregulation in male Japanese macaques provide an illustration of how intimately the causality of behaviour may be linked with physiology and morphology.

While it is not possible from this study to conclusively say whether the long-body structure of the younger males is the result of better nutrition in their new home, or primarily due to the acclimation effect of the new and distinctly hotter thermal regime, it is possible to show that the behaviour of the two body forms differ. To researchers who restrict themselves to the limited study of behaviour alone, without consideration of other possible causal factors, this result would be inexplicable.

While the study example has not dealt with the outer shell in terms of self-awareness, cognition, and tradition, it has provided an example of inferred feedback relationship between the environment, the behavioural interface, and the physiological/morphological structure of macaques. In the changed circumstances of the move to south Texas, the troop encountered a new environment with very different selective pressures influencing their survival. That they have survived and thrived is in part due to the efforts of their human caretakers over the years, but also due to their own system of response. Through shifts in behaviour, they were able to survive, then through deeper level changes in their ontogeny they have increased their range of behavioural flexibility and consequently thrived.

END NOTES

1. Behavioural Ecology is the discipline in which behavioural studies are related to aspects of the ecological circumstances. It focuses upon how behaviour, ecology, and evolution interact both at the proximate and ultimate levels of causality. This is concisely stated by Barnard (1983: 11) as "behaviour is the tool with which an animal uses its environment." The research perspective of the author is that of a behavioural ecologist.

2. Newton's Law of Cooling actually refers to a decrease in temperature rather than a loss of heat and in strict physiological terms may not be relevant to homethermic organisms. Kleiber (1961) notes:

> It is therefore unlikely that Newton's law of cooling is properly applicable to the heat loss of homeotherms where there is no decrease in temperature and therefore no cooling. The influence of body size on the rate of temperature loss (cooling) is different from the influence of body size on the rate of heat loss (without cooling). The application of Newton's law to the heat loss of homeotherms would imply that there is no such difference.
>
> The rate of heat loss is proportional to the surface area:

$$\frac{dQ}{dt''} = k_1 A(T_s - T_e).$$

the rate of heat loss is equal to the heat loss constant (k_1) times the surface area A, times the difference in temperature of surface (T_s) and temperature of the environment (T_e).

3. The Stefan-Boltzman law deals with the radiation emittance of a surface; whether that surface is organic or inorganic is immaterial. The relationship is described by Campbell (1977) in these terms:

> "The radiant energy emitted by unit area of surface of a blackbody radiator can be calculated from the Stefan-Boltzman law:
>
> $$\Phi_B = \sigma T^4$$
>
> where Φ_B is the emitted flux density (W/m^2), T is the Kelvin temperature, and σ is the Stefan-Boltzman constant ($5.67 \times 10^{-8}\ W\ m^{-2}\ K^{-4}$)."
>
> Which confirms that the radiative channel of heat transfer is also area related.

4. Fourier's law is more appropriate than Newton's for most physiological purposes since it concerns heat flow rather than temperature decrease. Kleiber (1961) provides the following:

Animal heat loss may be regarded as the flow of heat through an insulating surface layer, to which Fourier's law may be applied:

$$\frac{dQ}{dt} = \lambda \frac{A}{1} (T_R - Ts)$$

"where A = surface area, l = thermoconductive thickness of surface layer, λ = heat conductivity of surface layer, T_R = body temperature (rectal), Ts = surface temperature.".

CHAPTER SEVEN

Why Are Orangutans So Smart?: Ecological And Social Hypotheses

B. F. Galdikas and P. Vasey

Introduction

The exact role that intelligence played in hominid evolution has been questioned, but there can be little doubt that the original hominids were among the most intelligent of primates extant at that time. Research by Falk (1987) indicates that the brain size and morphology of the earliest hominids closely resembled that of extant pongids. In fact, no structure found in the human brain is absent from the brains of Old World anthropoids (Passingham, 1982). Because brain size can be used as a relative indicator of information-processing capacity (Jerison, 1983; 1985), it is probable that early hominid intelligence approximated that of extant pongids. We can assume that whatever form early hominid intelligence took, it has parallels in the extant pongids.

The earliest investigations of orangutans (Pongo pygamaeus) raised the question of why were orangutans so intelligent when they led such simple, solitary lives in the tropical rain forest. Researchers have commented on "The strange lackadasical way of life, here today, gone tomorrow...wandering and roaming through the jungles of Borneo..." (Reynolds, 1968: 157), that seemingly characterized wild orangutans. By contrast, orangutans in captivity have been taught to flake stone tools and have performed as well as chimpanzees in laboratory situations testing for intelligence and language acquisition (Wright, 1972; Beck, 1980; Lethmate, 1982; Miles, 1990). The question of why orangutans are so smart has not been put to rest despite numerous field studies that indicate orangutan lives in the wild are more complex than researchers have heretofore understood. Most researchers who conducted longer term orangutan studies (MacKinnon, 1978; Rodman, 1979; Galdikas, 1978b; also see Hrdy and Bennett, 1979) tried to explain orangutan intelligence on the basis of feeding ecology. Galdikas (1978b) went one step further arguing that orangutan intelligence was related to both feeding ecology and sociality. Within the broader context of the inquiry into the evolution of primate intelligence, researchers are still asking why orangutans possess considerable cognitive abilities (Dasser, 1985). If we can answer the question of why orangutans have advanced cognitive abilities, then perhaps we can begin to answer the question of why humans are so intelligent.

On the Pongid-Hominid Behavioural Repertoire

Hunting/meat eating, gathering, tool use, bipedality, language, food sharing, culture, inter-sexual ecological separation as reflected by a sexual division of labour and transportation of items to a central home base, have all been invoked as characteristics distinguishing hominids from pongids (Table #1). At one time or another, each of these behaviours has been cast as the "prime mover" responsible for pongid-hominid speciation (Washburn and Lancaster, 1968; Isaac, 1978; Zihlman and Tanner, 1978; Campbell, 1988). Survival on the open savanna, it is often argued, was dependent on expression of these "unique" hominid behaviours and natural selection favoured those transitional hominids who exhibited them frequently. As these distinguishing behaviours became part of the proto-hominid's daily behavioural pattern, they are thought to have acted as prezygotic isolation mechanisms (Mayr and Provine, 1980) which imposed barriers to gene flow with parapatric pongid populations. That is, these behaviours inhibited sexual interaction and exchange of genetic material between members of adjacent pongid and hominid populations.

The traits traditionally advanced as unique unto the hominids and as the "prime movers" of hominid evolution include:

Table #1:

Locomotory Adaptation:	-Bipedality
Social Behaviours:	-Food sharing
	-A sexual division of labour as reflected by ecological separation between the sexes
Foraging Behaviours:	-Hunting/meat eating/scavenging
	-Gathering
	-Transportation of food to a "Home base"
Technological Adaptations:	-Tool use/making
Cognitive Adaptations:	-Language capacities -Proto/culture

Beginning with Jane Goodall's (1963) observations of tool use and tool manufacture among chimpanzees in the wild, one by one the behaviours held as hallmarks of the hominid condition (Table #l) have fallen by the wayside as each has been found to be a component of wild pongid adaptations. Evidence now indicates that these behaviours constitute part of a shared pongid-hominid behavioural repertoire (Table #2). Thus, what separates the pongids from the hominids is a matter of degree rather than kind. Nevertheless, there is still a pervasive acceptance of these behaviours as crucial to hominid emergence. Implicit in this assumption is that these behaviours only gained major adaptive significance with the emergence of hominids and that they constitute irrelevant components of wild pongid adaptations.

Paleoanthropologists in general envision the origin and evolution of the behaviours traditionally described as the "prime movers" of hominid evolution as: [1] appearing at the point of pongid-hominid speciation, and [2] as products of direct selection due to extreme ecological pressures such as reduced tree cover, heat, aridity and predation, experienced in the ancestral savanna niche. Yet, given the existence of a shared pongid/hominid behavioural repertoire (Table #2) the origin of these "prime movers" of hominid evolution must have occurred long before pongid-hominid divergence (Table #2). Since this appears to be the case, a true understanding of the origin and evolution of these behaviours requires an examination of the pongids, as well as the hominids.

Table #2
Research Indicating at Shared Pongid/Hominid Behavioural Repertoire*

Behavioural Hallmark	Researcher (s)		Pongid
Hunting/Meat Eating:			
	Teleki	1975	Chimpanzee
	Sugardjito & Nurhuda	1981	Orangutan
	Tutin & Fernandez	1983	**Gorilla
	Badrain & Malenky	1984	Bonobo
	Takahata et al.	1984	Chimpanzee
	Goodall	1986	Chimpanzee
	Boesch & Boesch	1989	Chimpanzee
Gathering and Transport of Food:			
	Boesch & Boesch	1984a	Chimpanzee
	Chevalier-Skolnikoff et al.	1982	Orangutan
Tool Use/Making:			
	Goodall	1964, 1986	Chimpanzee
	Galdikas	1982	Orangutan
	Kano	1982	Bonobo
	Boesch & Boesch	1948b, 1990	Chimpanzee
	Wood	1984	Gorilla
	McGrew	1987	Chimpanzee
Proto/culture:			
	McGrew, Tutin & Baldwin	1979	Chimpanzee
	Nishida	1980	Chimpanzee
	Galdikas	1983	Orangutan
	Goodall	1986	Chimpanzee
	de Waal	1989	Bonobo
Language:			
	Shapiro	1982	Orangutan
	Patterson & Linden	1981	Gorilla
	Savage-Rumbaugh et al.	1989	Chimpanzee Bonobo
	Miles	1990	Orangutan

Food Sharing:	Teleki	1975	Chimpanzee
	Horr	1977	Orangutan
	Silk	1978	Chimpanzee
	Kuroda	1984	Bonobo
	Goodall	1986	Chimpanzee
	Bard	1990	Orangutan
	Hiraiwa-Hasegawa	1990	Chimpanzee
Ecological separation:	McGrew	1979	Chimpanzee
	Wrangham & Smuts	1980	Chimpanzee
	Galdikas & Teleki	1981	Orangutan & Chimpanzee
Bipedalism:	MacKinnon	1974	Orangutan
	Dixson	1981	Gorilla
	Susman	1982	Bonobo
	Bauer	1977	Chimpanzee
	Galdikas, pers. observ.		Orangutan

* Included are only a few of the mnay citations we could use to support our claim for a shared pongid/hominid repertoire.
** Although not meat eating *per se*, this article discusses a faunal component, termites, in the diet of one population of gorillas.

The behaviours traditionally advanced as hallmarks of the hominid condition are infrequently expressed by the pongids, which speaks strongly to the suggestion that the capacity for these behaviours was not the product of direct selection experienced in the ancestral hominid savanna niche. Had these behaviours been products of direct selection, wild pongids would manifest them more frequently. Why infrequently expressed behaviours evolved among the pongid, need not be a source of confusion once one considers that "Natural selection may build an organism 'for' a specific function or group of functions, [but] the 'purpose' need not fully specify the capacity of that organism. Objects designed for definite purposes can, as a result of their structural complexity, perform many other tasks as well." (Gould, 1980: 57). In short, "Structure has latent capacities." (Gould, 1980: 57). Along these lines, the pongid/hominid behavioural repertoire, as expressed by the great apes, represents a concomitant by-product and a consequence of directional selection for some other adaptive trait. Because the behaviours that consititute part of the shared pongid/hominid behavioural repertoire were not a product of directional selection, understanding their origin and evolution requires an appreciation of the basal adaptation which was being selected. It is this basal adaptation which laid the basis for the behaviours deemed the prime movers of hominid evolution.

We argue that the basal adaptation allowing for the shared pongid-hominid behavioural repertoire was, in large part, increased encephalization and, by extension, expanded cognitive abilities. Jerison (1983, 1985) suggests that encephalization, as measured by his encephalization quotient, represents the amount of additional brain mass an organism possesses above and beyond its basic somatic needs. Among the pongids, higher order cognition probably did not evolve as a specific ability, but rather as a more general adaptation for problem solving which was not domain specific. The evolution of a higher order, non-domain-specific intelligence would have allowed for increased information-processing abilities (Jerison 1983, 1985) and behavioural flexiblity (Napier, 1970; Parker, 1978; Burton, 1984), both of which would have been necessary prerequisites for expression of the behaviours constituting the shared pongid-hominid repertoire. In order to fully understand these behaviours exhibited by the great apes, yet hypothesized as crucial to the hominization process, we must first understand the origin and evolution of pongid intelligence. "To understand the origin of anything, we must have an overarching body of theory that governs both the thing itself and its precursor to successors" (Cartmill, 1990: 188).

Orangutans are the most phylogentically conservative of all the pongids. Unlike chimpanzees and mountain gorillas, they have never radiated out of the climax tropical forest where pongid intelligence intitially evolved. Our investigation into the origin and evolution of pongid intelligence will emphasize how orangutans utilize their intellectual capacities in the wild. In particular, we draw on data gathered on a population of wild orangutans at Tanjung Puting National Park, Indonesian Borneo over 1,273.5 hours from November 1971 to November 1975, as well as on more recent anecdotal data. This approach may serve to elucidate selective mechanisms that favoured the evolution of pongid intelligence in general, and by extension, which mechanisms laid the basis for the pongid-hominid behavioural repertoire.

Primatologists have long debated the relative importance of foraging problems versus social problems as selective mechanisms influencing the evolution of primate intelligence. In the following sections we address the role each may have played in the evolution of orangutan intelligence.

Feeding Ecology and the Rise of Orangutan Intelligence

During their waking hours, wild primates spend a disproportionate amount of time engaged in feeding activities. Because so much time is spent feeding and because food acquistion is intimately linked to survival and reproduction, primatologists began to investigate the role feeding ecology played as a selective mechanism mediating the evolution of primate intelligence. The basic premise underlying this theory is that as the foraging problems faced by a primate become increasingly complex, there is a greater need to deal with disparate types of information and to store and recall such information. In short, there is ever greater selection in favour of increased intelligence.

Because the majority of primates, including orangutans, feed at the first trophic level on plant material, their food acquistion behaviour does not involve the pursuit and subsequent kill of living, mobile prey. Instead, food acquistion by primates can be broken into three sequences: [1] food recognition, [2] search/foraging, and [3] handling/processing. Primatologists have questioned how each of these three feeding sequences might become increasingly complicated necessitating advanced intelligence and behavioural flexiblity. More specifically, dietary diversity as reflected by an opportunist/generalist adaptation (Parker, 1978; Galdikas, 1978b), the need to memorize and monitor food sources that are irregularly distributed in time and space (Wrangham, 1977; Galdikas, 1978b; Clutton-Brock and Harvey, 1980; Milton, 1981, 1988; Menzel and Juno, 1985) over a large supplying area (Menzel, 1978; Clutton-Brock and Harvey, 1980) and extractive foraging on embedded foods (Parker and Gibson, 1977, 1979; Gibson, 1986) have all been argued to place a high selective premium on intelligence. Add to this the need to decide when to leave a food patch given decreasing returns for energy and time expended, the need to avoid predators while foraging, and the need to acquire adequate food when faced with inter- and intra-specific competitors, and the challenge of feeding oneself becomes a formidable task.

Inferring the presence or absence of intellectual capacities among other species is difficult because intelligent behaviour as manifested among one species may be irrelevant behaviour to another. In an attempt to control for this problem, we refer to the following four criteria set out by Menzel and Wyers (1981) by which intelligent foraging behaviour might be empirically assessed:

[1] The ability to remember and to plan ahead,

[2] The ability to account for stimuli that are momentarily absent,

[3] The lack of a rigid one-to-one relationship between stimuli and responses (various behavioural patterns are be used to obtain the same goal),

[4] The ability to take into account a relatively large number of variables, rapidly and simultaneously.

i. **Recognizing Food**

The first step an orangutan faces in feeding itself is recognizing what to eat. To do so, inanimate objects in its range must be categorized as edible, nutritional, non-toxic and/or medicinal based on a variety of traits including size, texture, colour, and physical construction (criterion 3 and 4). In other words, a "taxonomy" of dietary and non-dietary items must be constructed (Burton, 1984: 77). Often, orangutans do not taste food prior to ingestion or rejection, but merely inspect it visually or handle and smell it. In the case of toxic food items, this behaviour suggests that the ingestion of toxins inducing nausea is not necessary for toxicity to be inferred (criterion 2).

If the need to recognize and choose appropriate food was an important selective mechanism influencing primate intelligence, dietary diversity as reflected by an opportunist/generalist adaptation should be positively correlated with brain size and intelligence (Parker, 1978). What evidence is there that this is the case among the large-brained, intelligent orangutans? With a diet characterized by a complex mix of fruit, leaves, bark, sap, insects, shoots/stems, honey and fungi and which varies markedly from month to month, orangutans can definitely be described as opportunistic foragers (Galdikas, 1988). Moreover, the dietary diversity exhibited by orangutans is remarkable among primates. At Tanjung Puting, 317 food types have already been identified. It is reasonable to assume that orangutans at this field site actually use more than 400 food types (Galdikas, 1988) and are capable of remembering and recognizing each of these (criterion 1).

In rainforest habitat, food availability regularly fluctuates (Sugardjito et al., 1987; Galdikas, 1988). The orangutans' ability to recognize a large number of food items greatly reduces their risk of starvation or malnurition during periods of food scarcity. Futhermore, food recognition reduces the possiblity of ingesting toxins. In the case of the opportunistic orangutans, it seems that dietary breadth played a significant evolutionary role in encephalization and intelligence. Where health and survival depend on obtaining nutrients from a variety of sources, most of which are seasonal, the ability to recognize, store and recall several criteria for each of the vast array of potential food items suggests considerable selective pressure.

ii. **Finding Food**

Once an orangutan recognizes appropriate food items, it must decide where to search, or forage, for them. Due to the dynamic nature of tropical forest fruit sources, complex decisions regarding the area in which to forage must continuously be made so that time and energy are not wasted randomly moving about and encountering unproductive food patches. Higher order intelligence allows orangutans to "cognitively map" (Menzel 1978: 378) irregularly distributed food patches within their home range, thereby permitting efficient foraging. Time and energy spent in travel are minimized by choosing the shortest routes between productive food sources. Among orangutans, such capacities are indicated by their use of remarkably efficient routes to reach out-of-sight (criterion 2), productive food patches (MacKinnon 1974, 1978; Rodman, 1979; Galdikas, 1978b). The ability to take "shortcuts" is perhaps the clearest single piece of evidence for cognitive mapping (Menzel, 1978). This behaviour entails the need for detailed individual memory of environmental features (criterion 1) and the ability to represent several such features in quantitative terms so that a least distance policy can be employed (criterion 1 and 4). Orangutans seemed to be better at this than primatologists equipped with compasses, maps and cut trails (Galdikas, 1978b). MacKinnon (1978: 121) notes:

> It is quite clear that orangutans do not amble aimlessly around their ranges taking pot luck. They move in a very purposeful and deliberate fashion. To plan such [economically efficient] routes they must not only have a very good knowledge of what is where, but they must also have a good sense of direction, distance, and travel time.

A least-distance policy during food aquisition has been similarly demonstrated for hamadryas baboons *(Papio hamadryas),* chimpanzees *(Pan troglodytes)* and saddle-backed tamarins *(Saguinus fuscicollis)* (Sigg and Stolba, 1981; Boesch and Boesch, 1984a; Menzel and Juno, 1985; Sigg, 1986).

During their study of hamadryas baboons Sigg and Stolba (1981; Sigg, 1986) attempted to find evidence for "cognitive mapping" other than the classic "shortcut." They hypothesized that the pattern of approach towards a "goal" should be distinct from the usual manner of travel if the baboons possessed cognitive maps. Along these lines, orangutans at Tanjung Puting were often observed to suddenly veer off from their previous route and travel more rapidly than usual just prior to making visual contact with their presumed "goal." Upon reaching the "goal" this noticeably rapid pattern of travel would cease. Movement of this sort suggests that orangutans cognitively map features of their environment and use this information to engage in purposeful, efficient travel (criterion 1) towards unseen (criterion 2) environmental features of their environment that constitute "goals."

Since orangutans do not move about their ranges randomly, could topographical features be directing their movements? At some of the mountainous sites where orangutans have been studied, steep slopes and deep valleys may direct orangutan movement in a non-random pattern that mimics the premeditated movement indicative of higher order intelligence. If this is the case, it would be difficult for the primatologist to infer cognitive mapping abilities and higher order intelligence. Such an explanation seems highly improbable for the Tanjung Puting orangutans, however, given the flat, featureless, uniform topography found at this site. It appeared that, irrespective of topography, these orangutans utilized arboreal pathways in novel combinations depending on the food source that constituted their "goal" (criterion 3 and 4).

Orangutans appeared to use their highly developed spatial memory to make systematic checks of food sources that had a high probabilty of being productive. In a single day, orangutans were able to find a number of productive stems of the same rare, irregularly distributed species. This suggests they may have recognized plant species that exhibit synchrony in food productivity and exploited this predictablity. In other words, when encountering one productive tree species, individuals may make systematic checks thoughout their ranges for other trees of that same species (criterion 1 and 2), in order to maximize their encounter rate with food. Similar abilities have been demonstrated in Japanese macaques *(Macaca fuscata)* (Menzel, 1991).

While orangutans appeared to remember the location of food sources and use this information to economize time and energy spent foraging, there was also evidence that orangutans monitored the phenological state of food sources even when such sources were not productive. Orangutans were frequently observed to enter trees, visually inspect the leaves, bark or fruit, discard the food item, and then exit the tree.

Because the orangutans did not eat any food on such occasions it appeared as if they were monitoring the state of the trees' productivity. In other words, an orangutan might observe the phenological state of a tree, conclude its fruit was far from ripe or its leaves too mature and toxic and then not return until some weeks later when fruit matured or a new flush of young leaves bloomed. Such behaviour involves remembering the location and phenological cycle of numerous food sources (criterion 1), assessing the phenological state of food sources based on several characteristics that change through time (e.g., color, size, physical construction, etc., of leaves, fruits, flowers, etc.) (criterion 3 and 4), and predicting the future productivity and potential of food sources based on their present state (criterion 2). The following example from Galdikas (1978b) serves to illustrate these points:

> ...an adult female, who in previous days had visited and exploited several durian trees, suddenly veered off into an area that to that time was outside her known home range. She travelled more rapidly than usual and barely foraged. During

> this rapid travel she passed through two different wild durian trees, neither of which was fruiting. After passing through [and inspecting] the second durian tree, the adult female returned to a part of her home range known to us. That she was inspecting the phenological state of the two durian trees is evidenced by the fact that, even though we eventually accumulated over one thousand hours of observation on this particular individual, we never again observed her passing within one hundred meters of either of those two specific durian trees.

We believe that the orangutans' ability to remember and monitor unproductive food sources without being cued by a productive source accounts in part, for their high intelligence and large brain/body ratio.

Although the evidence for orangutan cognitive mapping and monitoring of their environment is largely anecdotal, research carried out on captive saddle-back tamarins *(Saguinus fuscicollis)* under laboratory conditions indicates this tiny callitrichid utilizes cognitive maps to monitor food availablity at various productive and unproductive locations (Menzel and Juno, 1985). They do so through repeated movements back to previously productive food sources over extended periods of time. Orangutans seem to do exactly the same thing, over very short periods of time, utilizing several major food sources to which they repeatedly return in varying combinations on subsequent days. On the basis of these principles, it was sometimes possible to predict from the precise direction in which an orangutan was moving, the specific tree which provided the ultimate "goal," even though the tree might be over one kilometer away (criterion 1 and 2).

Relative to many species that feed at the first trophic level, the search efficency of orangutans is complicated because fruit, which is their dietary staple, is irregularly distributed both in time and space (Sugardjito et al., 1987; Galdikas, 1988). This means that fruit sources within an orangutan's home range are not only irregular in their location, but they are also irregular in their timing of productivity. In short, fruit is rarely an abundant resource in the tropical forest. It has been demonstrated among primates that frugivores, who exploit food that is irregularly distributed in time and space have larger brain/body ratios than folivores, who exploit food that is more uniform and dense in spatio-temporal distribution (Clutton-Brock and Harvey, 1980; Milton, 1981, 1988). Milton (1981) suggests that because leaves are denser and more uniform in time and space relative to fruit, folivores are not under intense selection pressure to remember the location and phenology of leaf sources. Instead, their foraging success appears dependent on very regular patterns of movement which maximize the probability of encountering leaves. This suggests that the need to remember and monitor the location and productivity of food sources that are irregular in their spatio-temporal distribution was a major selective mechanism influencing primate intelligence and encephalization (Clutton-Brock and Harvey, 1980; Milton, 1981, 1988).

Milton (1981, 1988) has argued that some degree of *predictablity is* associated with the spatio-temporal distribution of tropical plant foods, so that once the location of a food source is found it remains constant in space. Furthermore, she states that a measure of predictablity is also associated with the phenological pattern of each tree species and that this element of predictablity could be exploited by primates to "counteract" (Milton, 1981: 536) irregular productivity over time. In other words, primates possessing higher order intelligence could take note of a productive food source and make a systematic check of individual trees of the same species throughout their range. While we agree with Milton's (1981) hypothesis, the situation may be even more complicated for orangutans.

At Tanjung Puting, as at Milton's field site on Barro Colorado Island in Panama, fruit is much more patchy in its spatio-temporal distribution than are leaves (Galdikas, 1988) and, thus, orangutans are faced with a much more difficult task in locating and monitoring their dietary staple, relative to folivores. The flowering and fruiting patterns of individual species within the rainforest at Tanjung Puting do not follow predictable yearly cycles in the manner of temperate forest species. Each tree species seems to have its own flowering/fruiting cycle which is apparently triggered by very specific combinations of climatic conditions. Unlike at Milton's field site, however, the situation for a frugivore at Tanjung Puting becomes increasingly complex when one considers that tree species which do fruit in a given year vary greatly in the amount of fruit produced, even among trees of similar girth (Galdikas, 1978b; 1988). This can result in some trees within the *same* species producing no fruit while others are ladened with it. The availability of seasonal food resources is further affected by the "empty harvest" syndrome. This phenomenon occurs when the crop of a tree species appears abundant, but upon closer inspection the flesh of the fruit is either defective or not developed at all. Clearly, species synchronicity in productivity is by no means a uniform phenomenon at Tanjung Puting and, thus, the element of phenological predictablity which occurs at Milton's field site is a much less reliable guideline for these orangutans.

In sum, the phenological data from Tanjung Puting reveals a much more complex distribution of food over time than exists at Milton's (1981, 1988) field site in Panama. We believe the extremely complex and irregular distribution of food sources over time (and secondarily in space), coupled with conditions of poor visibility due to dense foliage, may have been major selective mechanisms favouring the evolution of intelligence and encephalization as exhibited by orangutans. These adaptations would allow for "cognitive mapping" of environmental features and the ability to monitor food sources regardless of their productivity state. These cognitive abilities allow orangutans to minimize travel costs while maximizing encounter rates with productive food sources.

iii. **Handling Food**

Once found, a food item may require handling, or processing, prior to ingestion in order to render it edible. Depending on the food type, processing can range from simply removing food from a branch to much more intricate manipulation of the food object. Food processing reaches its most complicated form with extractive foraging on embedded foods (Parker and Gibson, 1977, 1979; Gibson, 1986). Extractive foraging involves feeding on embedded foods which must be removed from the matrix in which they are encased or separated from inedible or toxic parts prior to consumption (criterion 2).

Paradoxically, while tool use by captive and ex-captive orangutans often surpasses that of the other pongids, both in quantity and variety, only the common chimpanzee *(Pan troglodytes)* has been observed to use tools to get food with any regularity in the wild (Wright, 1972; Beck, 1980; Galdikas, 1982; Lethmate, 1982; McGrew, 1989). Nonetheless, orangutans often process food items prior to ingestion (MacKinnon, 1974; Galdikas, 1978b, 1988; Rijksen, 1978; Chevalier-Skolnikoff et al., 1982; Wheatley, 1987).

At Tanjung Puting numerous food types and species, including embedded foods, were processed. Nuts and fruit encased in shell or husk were cracked open, inedible and toxic portions were detached from the edible portion of food items, the underground storage organs of plants were dug up, insects such as wood-boring termites were extracted as was honey contained in bees nests. The impressive size and toughness of many of the fruits that orangutans eat is not conveyed in most discussions of orangutan ecology. Wild durian *(Durio sp.),* for example, may weigh one kilogram and are difficult for a human to open without a machete. This is also true of such nuts as banitan *(M. leptopoda)* and pempaning *(Lithocarpus sp.)* to name only a few. When handling embedded foods orangutan processing behaviour was characterized by several complex, coordinated and simultaneous object manipulatory movements of several body parts (criterion 4) (Chevalier-Skolnikoff et al., 1982). Some foods such as bakunyit fruit were processed according to organized behavioural sequences which required successful completion of each sequence in the correct order to successfully render each item edible. The successful execution of such a behavioural sequence is based on a mental representation of conditions not present and the ability to plan ahead (criterion 1 and 2) (Chevalier-Skolnikoff et al., 1982). Attached substrates, such as level branches, were sometimes used as anvils while processing embedded foods. Food processing with the use of an attached substrate involved rapid and simultaneous coordination of substrate-food-body-force relationships (criterion 4) (Chevalier-Skolnikoff et al., 1982). The processing technique orangutans chose varied according to the food item being handled and some food items such as the burr-covered fruit *Chaeotocarpus castanocarpus* or bantian nuts *(Mezzetia leptopoda)* were processed using a variety of techniques (criterion 3). For example, *C. castanocarpus* was processed by rubbing handfuls of the fruit around a branch substrate, until all the burrs were removed; alternatively these fruits were processed by rubbing them on the back of the free hand (Galdikas, 1978b, 1982). The extremely hard-shelled bantian nuts were processed using three different techniques: [1] removal of the shell with the teeth only, [2] removal of the shell with the teeth as the mouth was covered with the back of the hand, the wrist or both hands or alternatively [3] removal of the shell as in [2] except in order to intensify the forced directed to crack the shell, the orangutan, while hugging a tree, pressed its cheeks against the tree trunk.

Prior to processing, orangutans commonly gathered a number of embedded food items and transported them to a specific place. This suggests that orangutans are capable of conceptualizing their impending food-processing behaviour and account for this by moving to a desired processing location (criterion 1 and 2). The transport of objects for the purposes of food processing has been similarly demonstrated among chimpanzees (Boesch and Boesch, 1984a). Possibly, orangutans were motivated to transport embedded foods to an appropriate substrate (anvil) so they could be processed more efficiently or they may simply have sought a more comfortable arboreal substrate on which to sit and feed.

Clearly, orangutans frequently demonstrate higher order intelligence while processing food, especially embedded foods. Parker and Gibson (1977, 1979; Gibson, 1986) have argued that radiation into a niche, where complex processing of embedded food items was necessessary prior to ingestion, would have placed a high selective premium on intelligence and encephalization. Selection for expanded inteligence and encephalization would have occurred in response to an increased need to solve the difficult problems embedded food processing presented. Gibson (1986) lends support to this hypothesis by demonstrating that, among primates, the largest brain/body ratios occur among omnivorous extractive foragers. We believe that the problems encountered during extractive foraging on embedded foods and during food processing in general were important selective mechanisms favouring orangutan intelligence. In particular, feeding on embedded foods would have promoted the need to represent the presence of an unseen food and then devise means to overcome physical and chemical defenses in order to access such food. Certainly, intelligent individuals in the ancestral Tanjung Puting orangutan population who could process food, particularly embedded food, and thereby render it edible, gained access to a wide range of dietary items that were unavailable to many other species. Preliminary observations, for example, indicate gibbons exploit some fruits in the immature state which orangutans eat primarily when these fruits are ripe. The mature fruits which gibbons do not exploit tend to be large, tough-skinned or deeply embedded inside a thick matrix.

iv. **Social Expertise and the Rise of Orangutan Intelligence:**

Students of primatology have marveled at the highly social nature of their subjects. Most primates spend many hours in proximity to one another traveling, feeding, playing, grooming, fighting, mating, and sleeping. They form life-long relationships that transcend generations and interact in complex ways according to their kinship, age, sex and individual history. Research suggests non-human primates perceive elements of their social systems (Cheney and Seyfarth, 1990) which are used to construct social "taxonomies" that inform the primate as to appropriate behaviour under specific circumstances (Burton, 1984: 77). Moreover, they use this social information to manipulate the behaviour of conspecifics for personal benefit. In this way primates are capable of using others as 'social tools' indicating that "primate intelligence is not just 'social,' but *Machiavellian* in its origins" (Whiten and Byrne, 1988a: 6). Because the social lives of most primates appear so complex, primatologists have suggested that the need to solve social problems was the major selective mechanism influencing primate intelligence and encephalization (Jolly, 1966; Burton, 1984; Cheney et al., 1986; Parker 1987; Byrne and Whiten, 1988a; Cheney and Seyfarth, 1990). More specifically, the ability to recognize one's self, other individuals and relationships, to participate in triadic interactions and to deceive, have all been cited as hallmarks of "Machiavellian" social intelligence (Whiten and Byrne, 1988b).

According to the Machiavellian social intelligence hypothesis, social situations are highly "reactive" necessitating that group-living primates continuously monitor the behaviour of conspecifics and accomodate their behaviour as the social milieu evolves (Whiten and Byrne, 1988a). Individuals capable of processing complex social information could manipulate the behaviour of conspecifics and influence the outcome of interactions for their own personal benefit. Those individuals able to attribute intentions, beliefs, knowledge and emotions to others rather than simply predict behaviours, would be far more adept in the social arena, because they could manipulate what another individual thinks rather than simply how it behaves (Cheney and Seyfarth, 1990). Attribution necessitates that the individual represent two, possibly contradictory, mental states at once. Social expertise of this magnitude would require an expanded ability to deal with disparate types of rapidly changing information and to store and recall such information. In short, faced with these types of social problems there would be ever greater selection in favour of increased intelligence and, by extension, encephalization.

Implicit in the social intelligence hypothesis is the idea that social complexity should be positively correlated with primate brain/body ratio when one has controlled for diet. Research by Sawaguchi (1990) on anthropoid primates lends support to this claim. Comparing the relative brain size (RBS) of congeneric species, he found the RBSs of polygamous multi-female/ multi-male species were larger than those of monogamous species. Extrapolating from this, one might predict that solitary primates would exhibit the lowest brain/body ratios of all the primates. Unfortuanately, Sawaguchi (1990) who classifies orangutans as solitary, does not elaborate on this point. We address the question of orangutans "solitariness" below and reassess this classification of their mating system.

Due to their enigmatic status as the "solitary" anthropoid, most researchers have tried to explain orangutan intelligence solely on the basis of feeding ecology (Mackinnon, 1978; Rodman, 1979; Hrdy and Bennett, 1979). Implicit in most considerations of orangutan intelligence is the assumption that they are socially retarded - exempt from all evolutionary pressures favouring a Machiavellian social intelligence. Rodman (1979) has argued that it would be difficult to make sense of orangutan intelligence on the basis of social complexity. Hence, in the context of the social intelligence hypothesis, the orangutan's high cognitive abilities and large brain/body ratio have always been regarded as problematic.

Investigations into orangutan sociality have been hampered, in part, because they frequently interact high in the canopy, shrouded by dense foliage and hidden from the scrutiny of any primatologist. Moreover, many orangutan researchers rarely observed interactions (Schaller, 1961; Yoshiba,

1964; Okano, 1965; Horr, 1975; Rodman, 1973), and often when interactions did occur the observer was unable to identify individuals. The complex task of addressing the role social pressures played in shaping orangutan intelligence is compounded by the fact that interactions between orangutans frequently involve only two independent individuals (although, dependent offspring may or may not be present). Two-thirds of the social interactions observed at Tanjung Puting, for example, involved only two independent individuals in contact. The absence of a third individual during social exchanges mitigates the potential for triadic interactions, the recognition of others' social bonds and the making of transitive inferences, all of which have been advanced as signifiers of higher social intelligence.

In the remainder of this chapter, we discuss evidence for orangutan possession of Machiavellian social intelligence. We restrict our discussion of social intelligence to wild orangutans whenever possible, however, some of the best evidence for Machiavellian social intelligence comes from observations of ex-captive orangutans. While we firmly believe that wild orangutans possess Machiavellian social intelligence, the nature of their semi-solitary adaptations largely masks its expression. Given the choice between utilizing their social intelligence or avoiding situations that require its expression, wild orangutans opt for the latter. In most circumstances social rewards hold little value for orangutans and thus, little incentive exists for the expression of social expertise. Orangutan social intelligence is most clearly demonstrated during interactions with primates for whom social rewards are great.

In this context, it is noteworthy that the ex-captive orangutans manifest the highest levels of social intelligence when interacting with humans, the Machiavallian manipulators par excellence. In order to clarify misconceptions regarding orangutan solitariness, our discussion of orangutan social intelligence will begin with a brief outline of the species social structure and mating system. We go on to address orangutan social intelligence from an egocentric perspective, examining their ability to recognize themselves, other individuals and their own personal relationships. We move out of this egocentric social realm to explore whether orangutans recognize others' relationships. This is followed by a discussion of how orangutans use their social knowledge which includes examples of coalition formation and deception. We submit that much of the evidence for wild orangutan social intelligence remains largely anecdotal, but console ourselves with the knowledge that, to date, this problem characterizes most studies of primate social intelligence in the wild (Byrne and Whiten, 1988b).

i. **Orangutan Sociality**

The basic units of orangutan populations have been shown to consist of: [1] lone adult males, [2] adult females usually accompanied by one or two dependent offspring, and [3] independent immature male and female individuals in transition between natal units and establishment of a separate solitary existence (Horr, 1975; MacKinnon, 1974; Rodman, 1973; Rijksen, 1978; Galdikas, 1979, 1985a). Although other researchers have reported the existence of transient females (MacKinnon, 1974; te Boekhorst et al., 1990) at Tanjung Puting all adult females and their dependent young occupied stable home ranges which overlapped extensively. By contrast, adult males demonstrated two patterns of ranging behaviour. Some adult males (residents) stayed in a limited home range for several years while others (transients) seemed to range more widely without staying in any one area for a prolonged period of time (Galdikas, 1979, 1985b; Mitani, 1985a; te Boekhorst et al., 1990). Exact figures pertaining to adult male home range size at Tanjung Puting are not available as none of the adult males remained in the study area throughout the four-year study period (Galdikas, 1979). Their movements indicated, however, that their ranges encompassed several of the adult females' home ranges, as has been reported at other sites (Rodman, 1973; Horr, 1975; Rijksen, 1978; Mitani, 1985b; Galdikas, 1979, 1988). Some subadult males appeared to adopt a transient ranging pattern, while other become residents (Galdikas, 1979, 1985c). Adolescent females often left the company of their mothers to range on their own both outside, but more commonly within their mothers' home ranges (Galdikas, 1978a). Eventually, adolescent females establish a home range adjacent to and overlapping their mothers' home range (Galdikas, 1978a).

Despite their seemingly asocial nature, orangutans do not exist in a social vacuum. Continuous, long-term research by Galdikas (1978a, 1979,

1981b, 1984, 1985a, b, c) has revealed the intricate web of social relationships which permeate wild orangutan social networks. Overall, orangutans at Tanjung Puting spent approximately 18.7% of their time in groups (Galdikas, 1985a). When considered separately, however, the various age/sex classes were participating differentially in social groupings (Galdikas, 1985a). The patterns of contacts exhibited a very definite asymmetry in sexual composition in that almost all groupings contained both males and females together or were exclusively female. Male-female contact, however, predominated.

At Tanjung Puting, adult males were totally intolerant of other adult males and were largely avoided by other orangutan age/sex classes. Outside of consortships, adult male orangutans were truly solitary, spending less than 2% of their time in contact with other individuals (Galdikas, 1985b). Solitary adult males used "long calls" to communicate with others and to mediate inter-male dominance hierarchies over long distances (Galdikas, 1983). By contrast, adult females responded to their own class in a much more variable manner and they participated in a variety of groupings outside of consortship. Focal adult females spent approximately 13.5% of the total time observed in social groups, often in the company of an adult female "friend" and her young (Galdikas, 1984). Mitani (1985a) observed repeated long associations between subadult males and adult females at his East Bornean study site. Thus, adult female orangutans are best described as semi-solitary (Galdikas, 1984).

Immature orangutans such as subadult males and adolescent females participated in a variety of groupings and the frequency and duration of such interaction was much higher than that of their adult counterparts (Galdikas, 1978a, 1985c). Moreover, immature were frequently responsible for initiating and maintaining contact with other orangutans for extended periods exceeding a week. Focal subadult males participated in groupings 40.9% of the time observed, whereas focal adolescent females were in groups 40.8 % of the time observed over and above the time they spent with their mothers. Thus, both are best described as social (Galdikas, 1978a, 1985c). Although orangutans sometimes exhibited overt affiliative gestures such as food sharing involving adult males and adolescent females, or grooming by adolescent females, orangutan sociality was more typically manifested in terms of sustained proximity during travel or rest, and by mutual tolerance during feeding.

ii. **The Orangutan Mating System**

Among primates, contributions towards offspring survival varies differentially depending on the parent's sex (Trivers, 1972). Although important exceptions exist, it is *usually* the primate female who raises and is in direct contact with the offspring for prolonged periods of time. Receptive orangutan females augmented this caregiving by mating with the fittest males, defined as mature males who have survived to adulthood, who could sustain a consortship, and often, who were dominant, resident individuals. Although males of some primate species participate extensively in the rearing of young, care of offspring by male orangutans, even in an indirect sense, was negligible (Galdikas, 1979). In absence of any role in offspring care, male orangutans directed such time and energy towards acquiring mating opportunities with females through intense male/male competition.

Extreme sexual dimorphism is often associated with polygamous mating systems (Selander, 1972) that are (usually) characterized by strong competition among the males (Mayr, 1972; Dunbar, 1988), be it for territories, places at a common breeding ground, or for females directly. Orangutans are among the most sexually dimorphic primates. Their social networks are structured in terms of dispersed, incomplete breeding units and *contra* Sawaguchi (1990) can be described as very loose fusion/fission, polygamous societies somewhat analogous to those of chimpanzees, yet in other ways resembling the mating system of gelada baboons *(Theropithecus gelada),* in which males usually do not interact cooperatively. Consequently, intense male/male competition and female choice are elements which typify orangutan sexual behaviour. The extreme level to which these behaviours have evolved among orangutans cannot be overemphasized. Some male orangutans, although capable of reproduction, probably never sire offspring.

Since orangutans are not group-living, the initial difficulty faced by each sex in terms of reproduction is locating a suitable partner. For adult males, the problem posed by female dispersal in space is compounded by the fact that female orangutans experience eight-year birth intervals (Galdikas and Wood, 1990). The rarity of receptive females both in time and space, increases male/male competition for mates. Galdikas (1983) has argued that one function of the adult male orangutan's long call is to advertise male presence and attract receptive females which males might otherwise experience difficulty finding. By long calling, however, males risk attracting potentially aggressive male competitors as well. Consequently, the calling rates of individual males was positively correlated with their rank. At Tanjung Puting subordinate males who called infrequently may have been less successful at locating receptive females and subsequently may have experienced lower lifetime reproductive success relative to calling males who could afford to advertise their presence.

Although Mitani (1985a, 1990) reports that adult males frequently copulated forceably at his fieldsite, forced copulation attempts by adult males were rarely observed at Tanjung Puting. Consequently, once a receptive female was found, an adult male had to maintain a consortship until ovulation occurred; this in the face of extreme aggression from other males. Intense competition among males for access to females was suggested by several factors, not the least of which was the adult males' total intolerance of one another. Aggression was invariably the consequence of direct adult male/male contact in the presence of adult females, while avoidance characterized lone male encounters (Galdikas, 1985c; Mitani, 1985a, 1990). Healed wounds and other physical anomalies indicated that some overt fighting occurred (Galdikas, 1985c). The extraordinary range of injuries and physical anomalies sported by most adult males suggested that direct male/male aggression must be far more prevalent than indicated by violent incidents observed.

Orangutan females were difficult for males to locate because they were silent and arboreal, slow moving and widely scattered in dense forest with limited visiblity. Unlike semi-terrestrial chimpanzees or gorillas, orangutan females could not be herded on "safari" or recruited. Hence, a female's movements could not be controlled by a male in the expectation that he would have access to her when she became fertile. If a male orangutan attempted to consort with an unwilling female, he could not guide her movements but instead had to follow her. Sometimes this path might lead straight to a dominant male whom she had chosen.

At Tanjung Puting, orangutans rarely congregated at food sources, so it was unlikely that males would encounter females by going to specific trees or vines. (This situation might be different in other areas where orangutans congregate in large fruiting fig trees.) Although some researchers question the importance of female choice as a selective mechanism influencing orangutan mating systems (Mitani, 1985b; Rodman and Mitani, 1987), many of the ingredients are present at Tanjung Puting for a mating system which is heavily structured by female choice. This system allowed adult females to monitor the male's physical fitness and to choose the best available mate.

Given the presence of female choice and male/male competition which characterizes the orangutan mating system, subadult males were in a particularly untenable position in regards to reproduction. Receptive females largely ignored subadult male sexual advances and instead traveled towards adult males long calling in the distance. Since they lacked the massive body size and strength which characterizes the adults of their sex, subadult males were inferior competitors when faced with adult male aggression. Consequently, risk of injury during such confrontations was high, and subadult males were always displaced (Galdikas, 1985c; Mitani, 1985a, 1990; Schurmann and van Hooff, 1986). Moreover, subadult males lacked the enlarged throat pouches of the adult males and as such, were unable to long call to attract receptive females. In an attempt to override the extreme levels of female choice and male/male competition which characterize orangutan mating systems, subadult male orangutans, unlike most other non-human primates, sometimes resorted to copulating forcibly with unwilling females (Galdikas, 1985c; Mitani, 1985a, 1990; Schurmann and van Hooff, 1986; Crawford and Galdikas, 1986). A female's resistance to such attacks ranged in intensity and duration from brief tussles with some squealing, pushing and slapping at the male's hands, to protracted, violent fights in which the female struggled throughout the length of the copulation, emitting loud "rape grunts" and biting the male whenever possible (Galdikas, 1985c; Mitani, 1985a). Even among subadult males, however, forced copulation was a secondary tactic, since subadults attempted (usually unsuccessfully) to maintain consortships with females whenever possible (Galdikas, 1985c; Schurmann and van Hooff, 1985; Crawford and Galdikas, 1986).

Like subadult males, receptive adolescent females were eager to consort with adults of the opposite sex (Galdikas, 1978a; Schurmann and van Hooff, 1985). They actively solicited adult males, who largely ignored or rebuffed their advances. Because female orangutans may exhibit a period of adolescent sterility, it may simply not have been worth an adult male's time and energy to consort, especially in the face of potential injury from aggressive male competitors. Hence, male mate choice played a large role in defining adolescent female sexual behaviour. Nonetheless, adolescent female orangutans were very persistent, at times slapping unreceptive adult males or pinching their genitals. Such persistence on the part of the adolescent female had its rewards, however. Aside from the obvious benefits of conceiving an offspring, by tenaciously consorting with an adult male, adolescent females established an affiliative bond that extended beyond consortship (Galdikas, 1978a), while gaining experience at consorting and protection from sub-adult males who might forcibly copulate with them.

Recently, Parker (1987) has reinvoked Charles Darwin's sexual selection model for the evolution of intelligence. She suggests that the most important social problems driving the evolution of higher intelligence was intra-sexual competition for suitable mates and epigamic selection (interactions between the sexes, such as mate choice). From Parker's (1987) sexual selection hypothesis, it seems reasonable to predict that as the intensity of intra-sexual competition and epigamic selection increase, so too do the cognitive abilities of the players. The strong elements of female choice and male/male competition among orangutans, coupled with the higher order intelligence demonstrated by this species, meshes well with these predictions.

Adult males must continuously monitor each other's spatial distribution through long calls. In addition, they must recall relationships with each other which are based on interactions as adolescents and exceedingly brief contacts as adults. At the same time, males must monitor the sexual receptivity of numerous females and demonstrate the social finesse to attract the attention of such females. In their turn, adult females must monitor the rank and physical condition of potential mates. Moreover, at any one time, females must assess their risk as potential victims of forced copulation attempts and then find means by which to evade such situations (a point we return to below in our discussion of deception). Subadult males must exercise their social expertise to convince receptive females to consort with them or, alternatively, to assess potential contexts in which forcibly copulating with a female carries with it acceptable levels of risk. The former is particularly tricky in the presence of an adult male consort, but not impossible (as we discuss later in the section on coalition). Adolescent females must be able to judge adult male behavioural states so that they can motivate males to consort, while at the same time not inciting aggression. Moreover, adolescent females must assess the fitness of subadult males who may become suitable consort partners when both are adults, while simultaneously avoiding forced copulation attempts.

In light of these difficulties, for an orangutan to achieve a successful reproductive career requires considerable social acumen. An individual's ability to avoid risk of attack from others, to assess the competitive ability of adversaries and to monitor the location, receptivity and fitness of potential mates determines his/her reproductive success. Social interactions such as these may involve at least three individuals. Triadic interactions are far greater in complexity than dyadic interactions because the need to process social information increases geometrically, rather than arithmetically (Carpenter, 1964; Harcourt, 1988). All this is no small feat given the highly dispersed nature of orangutan social networks. Knowledge of their social network and the individuals that form it would allow orangutans to consort and raise offspring successfully, in spite of the constraints imposed by their highly competitive mating system. We turn to these issues next.

iii. Orangutan Social Perceptivenes: Recognizing Oneself, Other Individuals And Personal Relationships

Although disagreement exists concerning the capacity of monkeys to recognize representations of themselves (Eglash and Snowdon, 1983; Anderson, 1984; Collinge, 1989; Cheney and Seyfarth, 1990; Burton, this volume), research has demonstrated that the great apes are able to recognize their own image (Gallup, 1982; Fossey, 1983). These observations indicate that the great apes may possess a concept of "self" as an individual, a capacity cited as evidence for Machiavellian social intelligence (Whiten and Byrne, 1988b). Beyond the realm of self-recognition, an ability to recognize the individuality of others would allow one to perceive personal relationships and the social elements, such as rank and kinship, that structure such relationships. Lack of this kind of social knowledge would prohibit the formation of life-long relationships and the potential for complex social interactions such as redirected aggression or alliance formation, which characterize primates.

Given that every orangutan is distinguished by its facial features, body build and unique personality, it is not surprising that orangutans recognized other members of their species as individuals as well as the personal relationships shared with such individuals. This is evidenced by the great disparity in reactions a single orangutan exhibited towards other members of its social network. Social relationships varied in nature from overtly affiliative to overtly aggressive depending on the individuals involved. For example, the adult female orangutan, Beth, ignored some males, ran away from others, and behaved affiliatively to a select few. Similary, Beth either fought, ignored or behaved affiliatively with females depending on the individual involved. One might argue that orangutans such as Beth do not recognize other members of their social network as individuals, but simply react in a different fashion upon each encounter, regardless of the individual involved. This conclusion seems inappropriate, however, since the nature of a relationship between two individuals was consistent over their lifespan. In other words, upon encountering each other in the forest, the reaction of two orangutans who shared a previous history was not random, but instead was predictable based on previous interactions. This was true even if individuals rarely interacted, indicating that orangutans must possess amazing social memory. Observations of one mother/offspring dyad at Tanjung Puting, for example, indicated that the female offspring travelled with her mother until she gave birth, at which time contact between the two greatly diminished. Yet, despite their apparent lack of interest, ten and a half years later, the mother and daughter were observed to move towards one another, embrace, and then travel together for four days (Galdikas, 1990).

A consideration of data pertaining to long-call behaviour indicates that adult males identified each other as individuals, assessed the other's rank relative to his own, and used this social information to behave appropriately. As mentioned, long calls function to mediate long-distance dominance relationships and spacing between adult males who rarely come into visual or physical contact with one another (Galdikas, 1983; Mitani, 1985b, 1990). If adult male orangutans were incapable of recognizing the calls identifying other adult males, one would expect: [1] the frequency of their calling to be unaffected by the calls of other males in the same area and [2] their reaction to the long calls of others to exhibit no particular pattern. The Tanjung Puting data indicates the contrary.

During the first four years of the Tanjung Puting study, three adult males: TP (Throatpouch), Nick, and Harry (HH) simultaneously shared at least part of the same home range within the main study area. All three males long called but did so at different frequencies (Galdikas, 1983). Nick called the most frequently, TP less so, and Harry called about half Nick's rate. Significantly, on days when Nick, TP or Harry were known to be within 800 meters of each other, Nick continued calling but TP and Harry only called at very low rates (Galdikas, 1983). These differences in call rates cannot be explained based on involuntary responses to random stimuli. Although some "sudden sounds" such as tree snags crashing to the ground provoked adult males to long call, 93.3% of all long calls heard seemed "spontaneous," in that they did not follow a sound cue (Galdikas, 1983). This indicates that orangutan males were long calling voluntarily, but did so based on the presence and proximity of other calling males in their vicinity (Galdikas, 1983). In other words, the identity of the audience affected the calling rate of a male.

Responses to the long calls of other males differed considerably depending on the caller's identity. Nick was never observed to move away from an adult male, no matter how close the male called to him. This contrasts with the two other males who, without making visual contact, ran away from Nick when he long called near them. Harry was also seen to run away upon hearing TP long call, yet TP did not react in a similar manner whenever Harry called. Once, upon hearing Nick's long call, Harry immediately sat upright in his night nest and urinated.

Because orangutans do not normally urinate in their nests, we assume that Harry did so in fear of possible contact with Nick. Dominance refers to the power one individual exerts over another through intimidation which may result in displacement of an opponent or priority of access to resources (Fedigan, 1982). It appears that a dominance hierarchy existed in which Harry was subordinate to both Nick and TP, and TP was subordinate to Nick (Galdikas, 1983). Since these incidents took place in different areas of the forest over a period of months, territorial impulses or varying emotional states cannot explain the reactions of individual males. Because adult males varied their calling frequency and their reactions to the long calls of others, it appeared that they used the long calls of other males to recognize individuals and form a mental representation of each other, in absence of visual contact. Possibly, the caller envisoned the effects of his call as other males moving away, while the receiving male envisioned the image of the calling male. Analysis of the acoustic properties of orangutan long calls by Galdikas and Insley (1988) revealed that caller identity could be ascertained based on the call's interpeak interval. Thus, orangutans may use interpeak interval to recognize the relationship between a particular call and a particular individual and that individual's rank relative to its own.

A primate capable of making transitive inference about rank relations, would be able to construct a dominance hierarchy based on partial information, without having to observe approach-withdrawal interactions between all possible dyads (Cheney and Seyfarth, 1990). At present, we do not have data which conclusively prove orangutans make transitive inferences regarding rank relations (e.g., if Nick is dominant over TP, and TP is dominant over Harry, is Nick dominant over Harry?). As we have stressed above, potential contexts for the type of observational learning necessary to make transitive inferences regarding rank relations were extremely limited for orangutans. This is especially the case for adult male orangutans since, during the first four years of the study, no instances were ever observed in which three adult males were simultaneously in visual contact. Hence, it seems unlikely that adult male orangutans construct dominance hierachies among themselves based on transitive inference. More likely, they rank each other on the basis of direct aggressive interactions. These conclusion are not intended to suggest that orangutans categorically cannot make transitive inferences. Given the laboratory evidence which suggests chimpanzees make transitive inferences (Gillan, 1981), our suspicion is that the other pongids, the gorilla and the orangutan, can do so as well. This is suggested by the behaviour of ex-captive orangutans during aggressive encounters. The attacked individuals do not run to humans with whom they have an affiliative relationship, but who are subordinate to the attackers. Instead, they often flee to particular humans with whom they share no affilliative relationship, but who are dominant over their attackers. Another provocative example involves a small, ex-captive subadult male who was trying to feed unsuccessfully due to the harassment of an ex-captive adult female (A). Suddenly, the subadult male noticed a second ex-captive adult female (B) who was dominant to the first female (A), moving towards the food. Even though the second female (B) was a hundred meters away, the subadult male suddenly relaxed his fear face as the first adult female (A) continued to harass him. He moved confidently to the food and began eating. Then, the first female (A) noticed the approach of the second female (B) and stopped harassing the subadult male. What was remarkable about this incident was the recognition by the subadult male that the arrival of the dominant female (B) ensured his access to the food, even while he was cringing under the attack of the first female (A). His glance and his change in behavioural state caused the first female (A) to alter her own behaviour. All three orangutans seemed to recognize the dominance relationships that existed and it remains possible that they did so using transitive inference.

iv. **Rocognizing and Comparing the Relationships of Others**

The social knowledge discussed in the previous section, is of a relatively simple order, as it concerns information that is gathered based on direct observation and personal experience. Taken alone, it would provide a flimsy argument for Machiavellian social intelligence among orangutans. Although orangutans spent a large proportion of their time alone, they were not confined within an egocentric social space. Instead, orangutans could climb outside the realm of their own personal relationships and recognize the relationships of others. Although it is not conclusive, the evidence indicates that they classify relationships into types, regardless of the individuals involved and, they make same/difference comparisons between types of relationships. Data on redirected aggression among orangutans at Tanjung Puting provide evidence that orangutans recognize kinship relations. During the first four years of the study, six acts of redirected aggression were observed, each involving an adult female attacking the juvenile or adolescent offspring of another adult female. The best documented of these involved associations between two adult females, Cara and Martha, and their respective offspring, the juvenile male Carl and the adolescent female Maud. Cara and Martha were stripping bark side by side on the trunk of a small tree, when Martha looked at Cara and made a motion to touch her. No aggressive intent was discernible, but Cara quickly looked up, backed away, left the tree, and then chased Martha's adolescent daughter Maud, in what seemed to constitute a classic case of re-directed aggression. A total of three of these attacks on Maud took place in Martha's presence, two attacks following Martha either touching Cara or initiating eye-to-eye contact as just described. A case might be made that Cara was simply venting frustration on a helpless target. This conclusion seems inappropriate, however, since Cara's son, Carl, was also present during these interactions and was often positioned closer to Cara than was Maud. Yet, Cara's aggression was not randomly distributed. She was never observed to attack her son Carl, in these situations, but instead moved past him to redirect her aggression at Maud. At other times, however, when they were alone, Cara *was* observed to attack Carl. It is also interesting that shortly after these attacks on Maud, Martha terminated travelling with Cara as if she herself had been attacked.

These observations suggest that orangutans perceive the kinship bonds of others in the sense that they recognize close associations between kin as qualitatively different to those shared by non-kin. It is difficult to explain acts of redirected aggression aimed at non-kin, without imputing intentionality to one's primate subjects (Hall, 1968; Menzel, 1978; Cheney and Seyfarth, 1990). For instance, it is possible that Cara recognized Martha "likes" Maud and by physically hurting Maud she could also affect Martha negatively; this would involve Cara attributing Martha's mental state and recognizing that Martha has emotions that she does not share. Furthermore, Cara's recognition of the bond shared between Martha and Maud would require that she compare Martha and Maud to each other, independent of herself, and draw the conclusion that they shared a social relationship analogous to the one she and Carl shared (e.g.,: Cara/Carl = Martha/ Maud).

Additional evidence that orangutans recognized and compared the relationships of others comes from observations of subadult males who followed behind consorting couples. The affiliative strength of a consort bond varied depending on the age of the female partner involved. Adult males actively sought adult females as sexual partners. By contrast they ignored or rebuffed many of the adolescent females' attempts to initiate consortships. When consorting with adult females, adult males were observed to cease termite foraging and terrestrial locomotion and to range shorter distances each day. There appeared to be no such compromise between consorting adult males and adolescent females since males continued to locomote on the ground and forage there for termites. Adolescent females experienced difficulty following males who moved rapidly along the ground. Furthermore, adolescent females were generally active for longer periods of time. Although adult males increased their day ranges somewhat, adolescent females sometimes continued to forage long after their male consorts had nested for the night. For these reasons, adult male/adolescent female consortships suffered from a lack of behavioural synchronicity that characterized consortships between adults.

Consorting couples were invariably followed by one or more subadult males (Galdikas, 1985c; Crawford and Galdikas, 1986). During the first four years of the study, subadult males attempted to "sneak" forced copulations with adolescent female consorts on three separate occasions (Galdikas, 1985c). Each of these occurred when the adult male moved away from the immediate vicinity of his adolescent consort, at which point the subadult male(s) which followed them raced towards the female. In one case, the adolescent female squealed, attracting the attention of her consort, who drove the subadult males away. Later, however, the adolescent female separated from her consort partner as he nested for the night. A subadult male seized the opportunity and forcibly mated with the adolescent, during which time a second, larger subadult male appeared, driving the first one away and forceably copulated with the adolescent female as well. During these attempts the adult male consort remained in his nest. This second subadult male was driven off the next morning when the adolescent female rejoined her adult male consort.

Subadult males were never observed to attempt sneak-forced copulations with adult female consorts. This was probably owing to the fact that adult males monitored their adult female consorts so closely that the chance a sneak-forced copulation would be successful was virtually zero (Crawford and Galdikas, 1986). These observations lead us to believe that sub-adult male orangutans *compared* the two types of consortships that exist (adult/adult versus adult/adolescent); they *recognized* the differences in affiliative strength inherent in each type of consortship, and, based on this social information, they *attempted* to sneak forced copulations only when a consort bond is relatively weak, such as those shared by adult males and adolescent females. If this interpretation is correct, this type of knowledge would allow subadult male orangutans to assess levels of personal risk associated with different social situations, which may result in an increased lifetime reproductive success.

v. **Putting Social Knowledge to Use: Coalitions.**

Triadic interactions in the form of alliances or coalitions are often cited as a measure of a higher order social intelligence among primates (Kummer, 1967; Harcourt, 1988; Whiten and Byrne, 1988b). Both these types of social interaction involve the solicitation of aid from a third individual against an adversary. Alliances involve long-standing social networks with kin and friends, while coalitions may arise due to more immediate considerations, such as temporarily perceived common interest (Fedigan, 1982). Frequently, these interactions arise in disputes over access to food and should be common when opportunties exist to monopolize food patches that are high in quality and clumped, but not very abundant (Harcourt, 1988).

Successful alliance or coalition formation involves complex decision-making processes dependent on the individual's pool of social knowledge. Harcourt (1988) argues that primate alliances are more complex and require greater information-processing ability than do the alliances exhibited by non-primate animals. Before entering into an alliance (or coalition), a primate must assess: [1] risk of injury to self from adversaries, [2] kin relationship to the adversary, [3] the competitive ability of the opponent and the opponents' supporters and [4] the nature of the relationship shared by potential allies and adversaries without reference to self (Harcourt, 1988). Moreover, a solicited individual must assess the risk of injury to the solicitor if denied support (Harcourt, 1988). Finally, as circumstances rapidly evolve, individuals must determine if the benefit of maintaining the alliance (or coalition) still offsets the potential cost of abandoning it or of injury. Individuals who cooperate in the form of coalitions or alliances to achieve common goals, possibly attribute intent to their partner(s). This involves recognizing that other individuals share a mental state amenable to her/his own (Cheney and Seyfarth, 1990).

Since orangutan food sources are dispersed, relatively rare, of high nutritional quality and clumped, the ecological ingredients exist to favour the existence of alliances. The dispersed nature of orangutan social networks, however, overwhelmingly mitigates their expression. Consequently, wild orangutans at Tanjung Puting have rarely been observed to enter into alliances, which leads us to believe that among orangutans, "coalition" may be a better term to describe this type of behaviour.

At Tanjung Puting two types of coalitions have been observed among wild orangutans over access to food patches. These include: [1] mother/adolescent daughter coalitions, and [2] adolescent female/subadult male coalitions against an adult female adversary. In each case, the coalition partners worked cooperatively to drive their adult female adversary out of a desired food patch. The benefit to forming such cooperative coalitions came in the form of immediate access to food. Normally, adolescent females deferred access to food patches that were occupied by an adult female, but by forming a coalition with either their mother or a subadult male "friend," they could displace adult females and take possession of a potentially rich food source. Sumatran adolescent female orangutans have been observed to form coalitions as well, in order to drive away siamangs that harrass them at food sites (Sugardjito et al., 1987).

The third type of coalition observed among wild orangutans occurred between subadult males against adult males consorting with adolescent females. During such interactions, subadult males appeared to act cooperatively to gain access to the adolescent female consort. (We stress that we are not attributing motivational states to the orangutans.) While one subadult distracted the adult male's attention, the other rushed towards the adolescent female and forcibly copulated with her. For subadult males, the benefit of engaging in such coalitions involved access to a potentially fertile adolescent female, with whom they could forcibly copulate. A particularly compelling example of risk assessment is offerred by an example of an ex-captive orangutan coalition. During this association, the risk of injury to the solicitor seemed to be assessed by the solicitee. This example involved a triadic interaction between one wild subadult male (A) and two ex-captive subadult males (B) and (C). The ex-captive "B" was dominant over "C" and both shared a relationship that could be described as affiliative. The incident began when "A" and "C" entered into an aggressive interaction, which "B" intently observed for a time. Following this, "B" rushed towards the pair (unbeknownst to "A"), and physically moved the subordinate sub-adult male "C" away from the area quickly. The dominant ex-captive "B" threated "A" with bared canines and continued to move "C" towards the observer. "B" exhibited behaviour typically associated with fear, and behaved as if he recognized the risk of injury to himself and to "C". As it turned out, "C" was badly wounded and would certainly have died had not a medical team operated on him the following day. After the operation, when "C" was put in a cage for his own protection, "A" appoached the cage, gazing at "C" as if he was checking "C's" condition. While these are only few examples of coalitions, they serve as reminders that orangutans are capable of processing large amounts of information pertinent to complex social interactions.

Deception

Deceptive behaviour provides excellent evidence for an ability to use social knowledge with such refined subtlety that victims are unaware of the fact they have been duped. Byrne and Whiten (1988b: 206) define deception as:

> (i) acts from the normal repertoire of an individual, (ii) used at low frequency, and in contexts different from those in which it uses the high frequency (honest) version of the act, (iii) such that another, familiar individual, (iv) is likely to misinterpret what the acts signify, (v) to the advantage of the actor.

Deceptive behaviour must occur at lower levels than the "honest" behavioural counterparts because, otherwise, the victims of deception would habituate to such acts and begin to question their validity. Because successful deception is frequency dependent, it usually occurs in the form of anecdotal data at levels that are not amenable for statistical testing (Byrne and Whiten, 1988b).

Most intraspecific acts of deception involve the manipulation of a target's attention or perception of reality (Burton, 1984; Whiten and Byrne, 1988c). Concealing oneself is a relatively simple form of deception which involves witholding information from another individual as opposed to active falsification of information. On occasion, orangutans deceived other individuals by concealing themselves so that the targets of deception were oblivious to the presence of the deceiver. Sitting motionless and/or in a concealed position behind vegetation rendered orangutans visually cryptic. To withold information in this manner may require that orangutans understand that others have different visual perspectives of their surroundings which limit their knowledge. If correct, this would mean that orangutans recognized the consequences of a visually exposed "self," the mental states of others and, that others may be ignorant of certain information. The following examples serve to illustrate these points.

On several occasions, the adult male, Harry, was observed to sit motionless in the canopy for over an hour after hearing other males (Nick and TP) long call at close range (Galdikas, 1983). Because the movement of an adult male orangutan through the canopy causes considerable noise, by behaving silently, Harry successfully deceived Nick and TP as to his location.

Harry's deceptive silence was much to his benefit, since both Nick and TP were dominant to him and, without doubt, would have behaved aggressively had either encountered him.

Like adult male orangutans, adult females sometimes concealed themselves behind foliage and remained still and silent. This behaviour was particularly evident when they perceived the presence of a sub-adult male in the area, probably because these males raise the potential threat of forced copulation. By silently concealing themselves, adult females evaded the unwanted attention of subadult males who, oblivous to their presence, would pass through the area uneventfully. Among adult females, this type of deception would be highly favored by natural selection because it would act as a countertactic against subadult male forced copulations, thereby allowing females to preserve their mate choice.

Although orangutans are capable of witholding information from others through concealment, they are also capable of actively falsifying information. This suggests that orangutans may recognize the existence of another individual's mental state and understand that this mental state may be false. One example of this type of deception involves the use of snag crashing (Galdikas, 1983) to "create an image" (Byrne and Whiten, 1988c: 213 Burton, this volume) that is threatening to the targets of deception. Snag crashing is an orangutan prototradition involving the detachment or disentanglement and throwing of dead tree snags (Galdikas, 1983). Among adult males, this behaviour probably functions as a prelude to an impending long call and to alert those within hearing distance as to the male's presence (Galdikas, 1983). Adult males who heard such crashes emanating from distances of less then 400m reacted strongly: they would long call, move away or remain motionless.

Adult female orangutans also snag crashed, but did so ten times less frequently than did adult males. Whereas snag crashing among adult males functioned as a prelude to a long call, snag crashing among adult females occurred at the approach of other orangutans. Because loud crashes, especially on windless days, frequently indicated the presence of another male in the vicinity, we suggest that adult female orangutans sometimes snag crashed to "create the image" of an adult male, possibly a consort, in their immediate vicinity. By doing so, they discouraged the company of other orangutans with whom they might have had to compete for precious resources. More interesting, however, is the possibilty that snag crashing by adult female orangutans represents a deceptive tactic to counteract the subadult males reproductive strategy of forced copulation.

Natural selection theory would predict the evolution of this behaviour among females, if it preserved the component of mate choice which subadults appropriate during forced copulations. Three lines of evidence lend support to this interpretation. Firstly, preliminary examination of female snag crashing indicates that approaching individuals moved away from the area the female snag crashed in, lending support to the hypothesis that females crash snag to "create the image" of an adult male in their immediate vicinity and deter others from interacting with them. Secondly, adult females appeared to use this tactic when the approaching orangutan did not witness them crashing the snag, suggesting that in the absence of visual information, the potential for misinterpretation of auditory (snag crash) information exists. Thirdly, it is unlikely that snag crashing by females was merely a manifestation of agonism since agonistic interactions between adult females and another orangutan unit frequently did not involve snag crashing. Because female snag crashing occurred at a much lower frequency than its "honest" counterpart (male snag crashing), the behaviour fits well with Whiten and Byrne's (1988b) definition of deceptive acts as rarely manifested, frequency-dependent behaviours.

Morphology normally constrains the possible range of deception of which an animal is capable. Nonetheless, additional evidence for deception among female orangutans involved one example of profound physiological changes in the genitalia which are usually associated with pregnancy (Crawford and Galdikas, 1986). Conception among female orangutans is marked by whitening and swelling of the female's genitalia. When a female's genitalia exhibit this condition, male orangutans appear less motivated to seek out their company and do not copulate with them. One free-ranging, ex-captive female who did not wander very far from camp, mated and consorted in turn with several wild adult males. When the female displayed swollen genitalia, the males immediately lost interest and did not congregate in her general vicinity. She did not become pregnant, however, until a year later when she again displayed swollen genitalia. Although this is only a single incidence, it is suggestive of the possiblity that female orangutans may deceive males as to their reproductive state. By inducing the genital swelling normally indicative of pregnancy females may falsify physiological information causing males to direct their sexual attentions eleswhere. This type of deception involves "creating an image" of pregnancy which sexually active males misinterpret.

The evolution of deception may be less constrained among orangutans than it is among other primate species for several reasons. Firstly, because orangutans are not group living and their survival is not dependent on cooperation, it is not imperative that signals be dispatched in an "honest" fashion. Therefore, orangutans may be freer than other primates to manipulate others using deceptive signals. Secondly, the risk of social retaliation against a deceptive orangutan who gets "caught," appears minimal since their lack of permanent social groups renders social punishment, such as ostracism, ineffective. Thirdly, the dense foliage of the tropical forest canopy often precludes verification of visual and auditory signals, thereby increasing the likelihood that a deceptive act will be successful. Fourthly, among orangutans the cost of being skeptical about signals may have evolved to prohibitive levels due to the intense male/male competition witnessed among this pongid. For example, males may not be able to chance the risk of (fatal) injury by being skeptical about a signal, such as a snag crash that is potentially the product of another more dominant adult male. For these reasons, the evolution of deceptive signalling may have been favoured among orangutans.

Conclusion

The behaviours traditionally held up as unique to hominids and as the "prime movers" of hominid evolution (Table #1), in fact, constitute part of a shared pongid/hominid behavioural repertoire (Table #2). Therefore, the capacity for the behaviours deemed hallmarks of the hominid condition must have evolved long before the origin of the hominids so that an examination of pongid adaptations may provide evidence concerning their origin and evolution. That wild pongids do not exhibit the behaviours in question on a habitual basis, strongly suggests that these behaviours were not the products of directional selection. Consequently, an examination of the basal adaptation which was directly selected and which provided the foundation for the pongid/hominid behavioural repertoire (Table #1) is required. We have argued that this basal adaptation was increased intelligence, or information processing ability, as reflected by an enlarged brain per body size.

For decades, primatologists have been engaged in debate concerning the relative importance of ecological problems versus social problems as selective mechanisms influencing the evolution of primate intelligence. We have attempted to address how each of these selective domains may have influenced the unique encephalization witnessed in the orangutan, and in the pongids in general. Orangutans exhibit higher order intelligence during all phases of the food acquisiton process: recognition of food, the search for food and food handling. A broad, opportunistic diet, irregular distribution of food in time and space and heavy reliance on embedded foods that require complex processing all function together to favour the orangutan's higher order intelligence within the ecological domain.

Because they are not group living, orangutan higher order intelligence has always seemed problematic when viewed from the perspective of the Machiavellian social intelligence hypothesis. Nevertheless, orangutans have not repudiated their anthropoid heritage and we argue that they possess (although rarely express) a higher order Machiavellian social intelligence. Research by Bard (1990) indicates that orangutans are capable of using gestures as "social tools" to manipulate others as early as three-and-a-half years of age. Orangutans are capable of recognizing themselves, other individuals, their own personal relationships and the relationships of others, and they put this social knowledge to use during acts of coalition and deception. This provides evidence (but does not prove) that orangutans are capable of recognizing: [1] the mental states of others, [2] that others may not be privy to the same information they are and [3] that the mental states of others may differ from their own. If this is true, it would mean orangutans could influence what others think and not just what others do.

In the past, primatologists have sought to disentangle the relative importance of ecological versus social variables as selective mechanisms favoring the evolution of primate intelligence (Hall, 1968; Cheney and Seyfarth, 1988, 1990). It is unlikely, however, that there is a simple and dichotomous answer to this question. Primate intelligence is the product of a complex and inseparable entwinement of both ecological and social mechanisms. Even among the relatively asocial orangutans, we believe social problems played an important role in shaping intelligence. Moreover, although we have focused here on feeding ecology and social interactions, these are not the sole mechanisms favouring orangutan intelligence. For example, Chevalier-Skolnikoff et al. (1982) argued that the contingencies of orangutan locomotory adaptation necessitate, and would have selected for, higher order intelligence. Enlarged body size probably played an important role in the evolution of pongid intelligence as well (Parker, 1978). Taken alone, the general trend towards an increase in body size set the stage for the increase in absolute brain size witnessed among the pongids. Coupled with the ecological and social selection pressures faced by pongids and discussed here, the trend towards a relative increase in brain size may have been magnified. Enlarged body size may have relaxed the threat of predation, thereby affording increased time to exercise curiosity about one's social and ecological world (Parker, 1978). Moreover, the strength afforded by enlarged body size may have permitted access to embedded food and enforced larger ranges which in turn, would have necessitated increased cognitive abilities. Home range is "presumably determined in part by how much spatial and object information can be assimilated" (Menzel, 1978: 398).

In conclusion, we have attempted to gain a deeper understanding of the hominid condition by stepping back and viewing ourselves as part of a long evolutionary history. The "...roots of uniquely human adaptations lie deep in the ecological adaptations of [our] arboreal frugivorous pongid ancestors" (Galdikas, 1981a), long before these ancestors adopted habitually terrestrial/bipedal modes of behaviour on the savannas. Therefore, a true understanding of the origin and evolution of the behaviours deemed hallmarks of the hominid condition requires an examination of the pongids as well as the hominids. In the end, we are left with the overwhelming impression that our hominid nature is but a profound expression of our pongid heritage.

ACKNOWLEDGEMENTS

The authors thank Dr. Frances Burton for her patience and encouragement throughout the preparation of this paper. Her comments proved to be most challenging and thought provoking and improved the paper immeasurably. Nancy Collinge and Alison Jolly gave critical thought and comment on various drafts. Murielle Nagy and Veronica Smith kindly provided technical assistance and Chris Knusel gave encouragement and help in locating references. Special thanks are due Dr. Richard Shutler and Ingrid Nystrom for their support. Paul Vasey extends thanks to the Department of Archaeology at Simon Fraser University for the 1991 Graduate Fellowship granted him.

Glossary: Definitions of Terms used in This Book

Agonism
From the greek, meaning the study of athletic combats; hence in ethology, the postures and rituals reflecting hostility between animals.

Alliance
An association between individuals which involves long-standing social networks.

Allogroom
Grooming someone else.

Allometric
Relative growth of one part of an organism in relation to the entire organism.

Allomothering
"Mothering" someone else.

Antagonism
Opposing, contestation, struggle against an adversary.

Anthropomorphic
Projecting human form - or metaphorically - human sensation, ideas, motives, onto non-human animals.

Behaviour
Minimally: anything an animal does; in psychology, actions of one organism which cause reactions in another. Behaviour may also be viewed as the organism's perceivable interface with its environment.

Behavioural Ecology
A study whose focus is the interaction of behaviour, ecology, and evolution at both the proximate and ultimate levels of causality.

Brachycephalic
Round headed.

Carnivore
A member of the Order including dogs and cats.

Cebus
A genus of new world monkeys.

Cercopithecine
Of or pertaining to the genus Cercopithecus.

Channel
Conduits of information; as in visual, auditory, olfactory, tactual.

Coalitions
An association between individuals which may arise due to immediate considerations, such as temporarily perceived common interest.

Coding
Putting information into a pattern.

Cognition
The act or ability of knowing, perceiving or conceiving, here defined as the capacity to think, to perceive, recognize and understand, is involved with the ability to selectively accept data input and to calculate proximate outcomes.

Consort
A constant companion, usually for quite limited times and especially during reproductive phases.

Conspecifics
Members of the same species.

Congenerics
Species of the same genus.

Consciousness
The state of being aware; the ability to monitor one's self, knowing one is doing that.

Contingency
A condition connected to another occurrence or event.

Demography
The study of the statistics of birth, death, disease, etc., with reference to a particular group.

Dolichocephalic
Narrow headed.

Dominance
Differential access to a limited, desired, resource.

Dynamic
The opposite of static changing, active, i.e., constantly in flux.
e.g. - exempli gratia, for example.

Empirically
Based on observation or experimentation. Hard evidence.

Epigamic selection
Interactions between the sexes, such as mate choice.

Epigensis
Originally suggested in the 17th century to oppose Preformationism, it is the biological finding that tissue differentiates in the dynamic process of development; it is not preordained or preformed.

Epiphenomenon
A trait which is the consequence of complex and interactive processes of development rather than traceable to gene action. Its current use describes a multi-factorial and multi-caused development. The concept eschews template models.

Epistemological
The method or theoretical basis for knowing.

Estrus cycle
Part of the reproductive cycle during which time the female is most receptive to the male and when she is most likely to conceive.

Existential
Concern with or pertaining to the individual's experience in the context of Existentialism.

Existentialism
Philosophy that the existence of the individual is free and the individual is responsible for his/her own development and being.

Fitness
Darwinian concept of differential contribution of one's genes to the next generation.

Frugivrorous
An organism whose diet emphasizes or consists mostly of fruit.

Genealogical
Of or pertaining to familial descent.

Genome
Genes considered as a group.

Heterogenistic
Constantly in dynamic flux and differentiation.

Hierarchy
A gradation of ranks.

Homeostasis
A concept introduced by Canon in 1932 as a general principle of physiology in which certain states are maintained at a constant level despite environmental fluctuation.

Homeotherm Animals
For example, mammals, that maintain their internal temperature at a constant value.

Homogeneity:
"...the most probable state of the universe or of an isolated system is random distribution of independent events, each having its own probability. In such a universe less probable states of nonrandom distribution tend to change to states of more random homogeneous distribution. The universe converges to homogeneity" (Maruyama, 1978:454).

Where parts of a system are all of the same kind; lack of differences.

Inclusive Fitness
The total contribution of one's genes through offspring and relatives.

Interpeak interval
In the study of communication, a quality of acoustics as is visualized on sonograms.

Kinesics
Body position.

Modality
Information is coded in a number of different ways which are generally called modalities. Modalities of the Visual Channel include coding of facial expression, hand gestures, etc.

M. fascicularis
The long-tailed macaque of s.e. Asia.

M. fuscata
The Japanese macaque.

M. mulatta
The rhesus macaque of India and China.

M. silenus
Lion-tailed macaque of south India.

M. sylvanus
Founder species of the genus Macaca, found in North Africa and the only European representative of an Old World monkey.

Matrilineal
Descent along the female line.

Metaethical
Meta = discourse above the level of inquiry.

Monogamy
Having but one spouse or mate.

Morphology
Form. The study of form.

Natal
Of or pertaining to birth, as in natal coat.

Nearest neighbour analysis
A statistical form of analysis to ascertain the significance of distance between individuals.

Nominal
Like a noun or name; existing in name only. Qualitative variables, like sex or religion or righthandedness, permitting comparison of similarity only.

Nomothetics
Of or pertaining to rules about phenomena.

Noyau
Living in a dispersed pattern in which mother-offspring bonds are the main social focus.

Oblate
Flattened at the poles.

Ontogenetic
Pertaining to ontogeny, the development of the individual.

Ontology
Study of the essence of being.

Ontological
Of or pertaining to the nature, essence, of something.

Ordinal
Defining something in a numerical series; second, third etc.; ordering points on a scale. In statistics, it refers to the third highest scale of measurement which can be compared to each other in terms of similarity and order only.

Parental investment
The contribution of a parent in an offspring, sometimes to the cost of another offspring.

Papio anubis
The common savanna baboon.

Papio hamadryas
The hamadryas baboon of Ethiopia.

Pan troglodytes
The common chimpanzee.

Parapatric
One of several terms used in evolutionary biology to describe the physical relationship between groups. Parapatric groups are adjacent to one another; stasipatric groups do not change their place.

Parsimonious
The notion in science that the greater the exactitude, the more precise and minimal the explanation for a given phenomenon, the greater the power of the explanation.

Pelage
Fur covering; usually used with reference to infants' natal coat.

Phenological
The timing of events in the life of a plant.

Phylogenetic
Pertaining to the evolutionary history of a phylum or group.

Poikilotherm Animals
Such as most reptiles, that depend on seeking appropriate sources of heat or cold (the sun; a burrow) to regulate their temperature.

Polygamy
Having more than one spouse or mate.

Pongo pygmaeus
The orangutan.

Prezygotic isolation mechanisms
One of several mechanisms which evolutionary biologists consider operate to keep species apart and integral.

Proceptive
The female is active in sexual encounters, rather than merely receptive or willing.

Prolate
Lengthened at the poles.

Proxemics
The study of spatial relationships.

Proximate
In the evolutionary study of behaviour: surface behaviours, like deception, which serve ultimate goals, i.e., reproductive fitness.

Proximity
Being near to another; nearness.

Quadrumanal
Having four hands; a four-limbed organism whose use of cheiridia (hands and feet) seems more significant as organs of manipulation rather than locomotion.

Qualitative
Descriptive; an analysis that focusses on the quality of things rather than their number.

Reproductive success
Fitness.

Reflexive
Looking or thinking back on itself.

Reification
Making real that which is only an idea.

Roles
Categories of behaviour necessary for the maintenance of the group but which do not have to be undertaken by all animals.

Socialization
The process of social maturation which implies the learning of group habits and traditions.

Sociobiology
The application of Natural Selection to social behaviour.

Socioecology
The evolutionary study of how ecological variables impinge on social organization.

Snag crashing
An orangutan prototradition involving the detachment or disentanglement and throwing of dead tree snags.

Stasis
Stoppage of flow, hence stationary; unchanging.

Stereotypic
Unchanging, or fixed; a behaviour or something which repeats without change.

Surface and deep structures
From linguistics, the theory that all languages have the same patterns at base, only the superficial aspects of statements vary.

Synchrony
Events occurring at the same time.

Sympatric
Two or more species of the same genus living in proximity.

Synchronicity
Events working at the same time as, or in conjunction with other events.

Synergism
Correlated action, especially between muscles, nerves, etc. Used here as cumulative interactive behavioural effects.

Taxis
Movement as a reaction to a stimulus.

Theropithecus gelada.
The gelada baboon.

Tradition
A local group's particular way of doing something; a pattern of behaviour specific to a local group, handed down generation to generation without language.

Triadic
Three way relationships.

Transcendent
Used here to mean extending beyond the domain of the individual.

Transitive inferences
The ability to infer the object or recipient of a behaviour. Aware of or implying the object or recipient of an action or behaviour.

Truncate
Cut off; shorten, taking the essential.

Ultimate
In the context of behaviour, those behaviours that have evolutionary implications.

REFERENCES

Abegglen, J. J. 1984. *On socialization in hamadryas baboons: A field study.* Bucknell University Press, Lewisburg.

Abitbol, M. M. 1988. Evolution of the ischial spine and the pelvic floor in the hominoidea. *American Journal of Physical Anthropology* **75**: 53-67.

Alcock, J. 1975. *Animal behavior: An evolutionary approach.* Sinauer Association Inc, Sunderland, Mass.

Aldrich-Blake, F. G. 1970. Problems of social structure in forest monkeys. J. H. Crook (ed.), *Social behaviour in birds and mammals.* Academic Press, London. Pp. 70-101.

Alexander, B. K. 1970. Parental behaviour of adult male Japanese monkeys. *Behaviour* **36**: 270-285.

Alexander, R. D. 1974. The evolution of social behaviour. *Annual Review of Ecology and Systematics* **5**: 324-382.

Alexander, R. McNeil. 1985. Body size and limb design in primates and other mammals. In W. L. Jungers, (ed.), *Size and scaling in primate biology.* Plenum Press, New York. Pp. 337-343.

Allee, W. C. 1931. *Animal aggregation: A study in general sociology.* University of Chicago Press, Chicago.

___________. 1931. *The Social life of animals. W. W.* Norton, New York.

Allen, J. A. 1877. The Influence of physical conditions in the genesis of species. *Radical Review* **1**: 108-140.

Allen, M. L. and Lemmon, W. B. 1981. Orgasm in female primates. *American Journal of Primatology* **1**: 15-34.

Altmann, J. 1974. Observational study of behavior: Sampling methods. *Behaviour* **49**: 227-267.

___________. 1980. *Baboon mothers and infants.* Harvard University Press, Cambridge.

Altmann, S. A. 1974. Baboons, space, time, and energy. *American Zoologist* **14**: 221-248.

___________. 1967. *Social communication among primates.* University of Chicago Press, Chicago.

___________. and Altmann, J. 1970. *Baboon ecology.* University of Chicago Press, Chicago.

___________. 1979. Demographic constraints on behaviour and social organization. In I. S. Bernstein and E. O. Smith (eds.), *Primate ecology and human origins.* Garland, New York. Pp. 47-62.

Andelman, S. J. 1986. Ecological and social determinants of Cercopithecine mating patterns. In R. W. Wrangham and R. Rubenstein (eds.), *Ecological aspects of social evolution.* Princeton University Press: Princeton, N.J. Pp. 201-216.

Anderson, C. 1986. Predation and primate evolution. *Primates* **27**: 15-39.

Anderson, J. R. 1983. *The Architecture of cognition.* Harvard University Press, Cambridge, Mass.

___________. 1984. Monkeys and mirrors: Some questions for primate psychology. *International Journal of Primatology* **5**: 81-98.

Andrews, M. W. 1986. Contrasting approaches to spatially distributed resources by Saimiri and Callicebus. In J. G. Else and P. C. Lee (eds.), *Primate ontogeny, cognition and social behaviour.* Cambridge University Press, Cambridge, Mass.

Argyle, M. 1988. *The Psychology of interpersonal behavior.* 4th Ed. Harmondsworth Middlesex, Penguin Books, Middlesex.

___________. 1978. Sequences in social behaviour as a function of the situation. In G. P. Ginsburg (ed.), *Emerging strategies* in *social psychological research.* John Wiley, New York. Pp. 11-37.

Asquith, P. J. 1986. Anthropomorphism and the japanese and western traditions in primatology. In J. G. Else, and P. C. Lee, (eds.), *Primate ontogeny cognition* and *social behaviour* Cambridge University Press, Cambridge. Pp. 61-71.

___________. 1984. The Inevitability and utility of anthropomorphism in description of primate behaviour. In R. Harr and V. Reynolds (eds.), *The meaning of primate signals.* Cambridge University Press, Cambridge. Pp. 138-176.

Bachmann, C. and Kummer, H. 1980. Male assessment of female choice in Hamadryas baboons. *Behaviour, Ecology and Sociobiology* **6**: 315-321.

Badrain, N. & Malenky, R. K. 1984. Feeding ecology of *Pan paniscus* in the Lomako Forest, Zaire. In R. L. Susman (ed.), *The pygmy chimpanzee: evolutionary biology and behaviour.* Plenum Press, New York. Pp. 275-299.

Baldwin, J. M. 1902. *Development and evolution.* New York.

Bard, K. A. 1990. Social tool use by free-ranging Orangutans: A Piagetian and developmental perspective on the manipulation of an animate object. In S. T. Parker and K. R. Gibson (eds.), *Language and intelligence in monkeys and apes: Comparative developmental perspectives.* Pp. 356-378.

Barnard, C. J. 1980. Flock feeding and time budgets in the house sparrow *(Passer domesticus). Animal Behaviour* **28**: 295-309.

Basmajian, J. V. 1967. Control of individual motor units. *American Journal of Physical Medicine* **46**: 480-486.

___________. 1963a. Control and training of individual motor units. *Science* **141**: 440-441.

___________. 1963b. Conscious control of single nerve cells. *New Scientist* **20**: 662-664.

Bauer, H. R. 1977. Chimp bipedal locomotion in the Gombe National Park. *Primates* **18**: 913-921.

Baulu, J. and Redmond, D. E., Jr. 1978. Some sampling considerations in the quantitation of monkey behavior under field and captive conditions. *Primates* **19**(2): 391-400.

Beck, B. 1980. *Animal tool behaviour: the use and manufacture of tools by animals.* Garland, New York.

Benedict, B. 1969. Role analysis in animals and men. *Man (n.s.)* **4**: 203-214.

Bergmann, H. 1847. *Ueber die Vernhltnisse der Wrmeoknomie der Tiere zu Ihrer Grosse.* Gottigen.

Bernstein, I. S. 1971. Agonistic behaviour during group formation and interactions in sooty mangabeys. *Proceedings 3rd International Congress of Primatology* **3**: 66-70. Basel, Karger.

___________. 1966. Analysis of a key role in a Capuchin (Cebus *albifrons)* group. *Tulane Studies in Zoology* **13**: 49-54.

Bernstein, I. S. and Sharpe, L. G. 1966. Social roles in a rhesus monkey group. *Behaviour* **26**: 91-104.

___________ Gordon, T. P. and Rose, R. M. 1974. Aggression and social controls in Rhesus Monkey *(Macaca mulatta)* groups revealed in group formation studies. *Folia Primatologica* **21**: 81-107.

Bertram, B. C. R. 1976. Kin selection in lions and evolution. In P. P. G. Bateson and R. A. Hinde (eds.), *Growing points in ethology.* Cambridge, Cambridge University Press. Pp. 281-301.

___________. 1978. Living in groups: predators and prey. In J. R. Krebs and N. B. Davies (eds.), *Behavioural ecology: An evolutionary approach.* Sunderland, Mass., Sinauer. Pp. 64-96.

___________. 1980. Vigilance and group size in ostriches. *Animal Behaviour* **28**: 278-286.

Bertrand, M. 1969. *The Behavioural repertoire of the stumptail macaque:* A *descriptive and comparative study.* Basel, Karger.

Boesch, C. & Boesch, H. 1990. Tool use and tool making in wild Chimpanzees. *Folia Primatologica* **54**: 86-99.

___________. 1984a. Mental map in wild Chimpanzees: An analysis of hammer transports for nut cracking. *Primates* **25**(2): 160-170.

___________. 1984b. Possible causes of sex differences in the use of natural hammers by wild Chimpanzees. *Journal of Human Evolution* **1**: 415-440.

___________. 1989. Hunting behavior of wild Chimpanzees in the Ta National Park. *American Journal of Physical Anthropology* **78**: 547-573.

Booth, C. 1962. Some observations on behaviour of Cercopithecus Monkeys. In J. Buettner-Janusch (ed.), *The relatives of man.* New York Academy of Science, New York. **102**: 477-487.

Box, H. O. 1980. *Primate Behaviour and Social Ecology.* Chapman and Hall, London.

Braestrup, F. W. 1971. The Evolutionary significance of learning. *Vidensk. Meddr. dansk naturh. Foren.* **134**: 89-102.

Bramblett, C. A. 1976. *Patterns of primate behavior.* Mayfield Publishing Company, Palo Alto, California.

___________. 1989. Mental well-being in anthropoids. In E. F. Segal (ed.), *Housing, care, and psychological wellbeing of captive and laboratory primates.* Noyes Publications. Pp. 1-11.

Brody, S. 1964. *Bioenergetics and growth: With special reference to the efficiency complex in domestic animals* [1945]. Reinhold Publishing, Hafner, New York.

Bruner, J. S, Jolly, A. and Sylva, K. (eds.). 1976. *Play: its role in development and evolution.* Basic Books, New York.

Burton, F. D. 1971. Sexual climax in female *Macaca mulatta*. In *Proceedings Third International Congress of Primatology*. S. Karger, Basel. **3**: 180-191.

___________. 1972. The Integration of biology and behavior in the socialization of *Macaca sylvana* of Gibralter. In F. E. Poirier (ed.), *Primate socialization*. Random House, New York. Pp. 29-62.

___________. 1977. Ethology and the development of sex and gender identity in Non-Human Primates. *Acta Biotheoretica* **26** (1): 1-18.

___________. 1984. Inferences of cognitive abilities in Old World Monkeys. *Semiotica* **50**(1-2): 69-81.

___________. and Bick, M. J. A. 1972. A Drift in time can define a deme: The Implications of tradition drift in Primate societies for Hominid evolution. *The Journal of Human Evolution* **1**: 53-59.

___________. and Chan, L. K-W. 1987. Interspecific babysitting. *Canadian Journal of Zoology* **65**: 752-755 .

Byrne, R. and Whiten, A.. (eds.). 1988a. *Machiavellian intelligence: Social expertise and the evolution of intellect in monkeys, apes and humans*. Oxford University Press, Oxford.

___________. 1988b. Tactical deception of familar individuals in baboons. In R. Byrne and A. Whiten, (eds.), *Machiavellian intelligence: Social expertise and the evolution of intellect in monkeys, apes and humans*. Oxford University Press, Oxford. Pp. 1-9, 205-210.

___________. 1988c. Tactical deception in primates. *Behavioural and Brain Sciences* **11**: 233-273.

Campbell, B. 1988. *Humankind emerging*. [5th ed.]. Little, Brown, Boston.

Campbell, G. S. 1977. *An Introduction to environmental biophysics*. Springer Verlag, New York.

Candland, D. K., and Kyes, R. C. 1986. Introduction: The Human primates' theory of the Primate mind. In J. G. Else, and P. C. Lee (eds.), *Primate ontogeny, cognition and social behaviour*. Cambridge University Press, Cambridge. p. 1

Carpenter, C. R. 1964. *Naturalistic behavior of Non-human Primates*. Pennsylvania State University Press, University Park.

___________. 1942. Societies of Monkeys and Apes. *Biological Symposia*. **8**: 177-204.

Cartmill, M. 1990. Human uniqueness and theoretical content in paleoanthropology. *International Journal of Primatology* **11**(3): 173-192.

Chadwick-Jones, J. K. 1991. The Social contingency model and olive baboons. *International Journal of Primatology* **12**(2): 145-161.

__________. 1989. Presenting and mounting in non-human primates: Theoretical developments. *Journal of Biological and Social Structures* **12**: 319-333.

__________. 1987a. A Social psychology of non-human primates: A developing field. *New Ideas in Psychology* **5**(1) 111-115.

__________. 1987b. Social psychology and primatology: Proximate explanations. *Ethology* **74**(2): 164-169.

__________. 1986. Social exchange, social psychology and economics. In A. McFadyen and H. MacFadyen (eds.), *Economic psychology.* New Holland, Amsterdam. Pp. 249-268.

Chance, M. R. A. and Mead, A. P. 1953. Social Behaviours and Primate Evolution. *Experimental Biology, Evolution* **7**: 359-439.

Chapais, B. 1983a. Dominance, relatedness, and the structure of female relationships in rhesus monkeys. In R. A. Hinde (ed.), *Primate social relationships: An integrated approach.* Blackwell, Oxford. Pp. 209-219.

Cheney, D. L. and Seyfarth, R. M. 1990. *How monkeys see the world.* University of Chicago Press, Chicago.

__________. 1988. Social and non-social knowledge in Vervet Monkeys. In: R. Byrne and A. Whiten (eds.), *Machiavellian intelligence: Social expertise and the evolution of intellect in monkeys, apes and humans.* Oxford University Press, Oxford. Pp. 255-270.

__________. 1985. Vervet Monkey alarm calls: Manipulation through shared information? *Behaviour* **94**: 150-166.

__________. 1982. Recognition of individuals within and between groups of free ranging vervet monkeys. *American Zoology* **22**: 519-529.

Cheney, D. L., Seyfarth, R.M. and Smuts, B. B. 1986. Social relationships and social cognition in Non-Human Primates. *Science* **234**: 1361-1366.

Chepko-Sade, B. D. and Sade, D. S. 1979. Patterns of group-splitting within matrilineal kinship groups. *Behavioral Ecology and Sociobiology* **5**: 67-86.

Chevalier-Skolnikoff, S., Galdikas, B.M.F. and Skolnikoff, A. Z. 1982. The Adaptive significance of higher intelligence in wild orangutans: A preliminary report. *Journal of Human Evolution* **11**: 639-652.

Chism, J. B. and Rowell, T. E. 1986. Mating and residence pattern in male patas monkeys. *Ethology* **72**: 31-39.

Clutton-Brock, T. H. and Harvey, P. H. 1980. Primates, brains and ecology. *Journal of Zoology of London* **190**: 309-321.

___________. and Harvey, P. H. 1976. Evolutionary rules and primate societies. In P. P. G. Bateson and R. A. Hinde (eds.,) *Growing Points in Ethology*. Cambridge, Cambridge University Press. Pp. 195-237.

___________. 1977. Primate ecology and social organization. *Journal of Zoology* **183**: 1-39.

___________. 1977. *Primate Ecology*. New York, Academic Press.

___________. Guiness, F. E. and Albon, S. D. 1982. *Red deer: The behaviour and ecology of two sexes.* University of Chicago Press, Chicago.

Cohen, J. E. 1969. Natural primate troops and a stochastic population model. *American Naturalist* **103**: 455-477.

Collinge, N. E. 1989. Mirror reactions in a zoo colony of cebus monkeys. *ZooBiology* **8**: 89-98.

Colmenares, F. 1990. Greeting behaviour in male baboons, in communication, reciprocity and symmetry. *Behaviour* **113**: 81-116.

Connell, J. H. 1984. On the prevalence and relative importance of interspecific competition: Evidence from field experiments. In G. W. Salt (ed.), *Ecology and evolutionary biology, a round table on research.* University of Chicago Press, Chicago. Pp. 661-696.

Crawford, C. and Galdikas, B. M. F. 1986. Rape in non-human animals: An evolutionary perspective. *Canadian Psychology* **27**(3): 215-230.

Crook, J. 1989. Socioecological paradigms, evolution and history: perspectives for the 1990s. In V. Standen and R. A. Foley (eds.), *Comparative socioecology: The behavioural ecology of humans and other Mammals.* Oxford, Blackwell Scientific Publications. Pp. 1-36.

Crook, J. H. 1964. The evolution of social organization and visual communication in weaverbirds *(Ploceinae). Behaviour (Supplement)* **10**: 1-178.

__________. 1970. The socio-ecology of primates. In J. H. Crook (ed.), *Social behaviour in birds and mammals.* Academic Press, London. Pp. 103-166.

__________. Ellis, J. E., Goss-Custard, J.D.. 1976. Mammalian social system: structure and function. *Animal Behaviour* **24:** 261-274.

__________. Gartlan, J. S. 1966. Evolution of primate societies. *Nature* **210**: 1200-1203.

Dahl, J. F. and Smith, E. O. 1982. Thermoregulatory, microhabitat preferences of stumptail macaques (Abstract). *American Journal of Physical Anthropology* **57**(2): 179.

__________. Berenstein, Williams, I. S. and L. 1986. Thermoregulation and social structure of a captive group of rhesus macaques. In D. Taub and F. King (eds.), *Current perspectives in primate social dynamics.* Van Nostrand, New York. Pp. 243-256.

Dasser, V. 1985. Social and non-social world of primates: Discussion section. In R. A. Hinde, A-N Perret-Clermont and J. Stevenson (eds.), *Social relationships and cognitive development.* Oxford Clarendon Press, New York. Pp.

Davenport, R. 1967. The Orang-utan in Sabah. *Folia Primatologica* **5**: 247-263.

Dawkins R. 1976. *The Selfish gene.* Oxford University Press, Oxford. Social relationships and cognitive development. Clarendon Press, Oxford.

__________. 1987. *The Blind watchmaker.* W. W. Norton, New York.

Deacon, T. W. 1990. Problems of ontogeny and phylogeny in brain-size evolution. *International Journal of Primatology* **11**(3): 237-282.

Deag, J. M. and Crook, J. H. 1971. Social behaviour and 'agonistic' buffering in the wild barbary macaque *(Macaca sylvanus). Folia Primatologica* **15**: 183-200.

Degabriele, R. 1980. The physiology of the koala. *Scientific American* **243**: 94-99.

Degabriele, R. and Dawson, T. J. 1979. Metabolism and heat balance in an arboreal marsupial, the koala *(Phascolarctos cinereus). Journal of Comparative Physiology* **34**: 293-301.

Deputte, B. C. 1982. Duetting in male and female songs of the White-cheeked Gibbon *(Hylobates concolor leucegenys).* In C. T. Snowdon, C. H. Brown and M. R. Peterson (eds.), *Primate communication.* Cambridge University Press, Cambridge. Pp. 67-93.

DeRousseau, C. J. and Reichs, K. J. 1987. Ontogenetic plasticity in non-human Primates: 1. Secular trends in the Cayo Santiago macaques. *American Journal of Physical Anthropology* **73**: 279-287.

De Ruiter, J. 1986. The Influence of group size on predator scanning and foraging behaviour of wedgecapped capuchin monkeys (*Cebus olivaceus*). *Behaviour* **98**: 240-258.

Devore, I. 1963. A comparison of the ecology and behaviour of monkeys and apes. In S. Washburn (ed.), *Classification and human evolution.* Aldine, Chicago. Pp. 301-319.

___________. 1965. Male dominance and mating behaviour in baboons. In F. A. Beach (ed.), *Sex and behavior.* Wiley: New York. Pp. 305-335.

___________. and Hall, K. R. L.. 1965. Baboon ecology. In I. Devore (ed.), *Primate Behaviour: Field studies of monkeys and apes.* Holt, Rinehart, Winston, New York. Pp. 20-52.

Dittus, W. P. J. 1975. Population dynamics of the toque monkeys, *Macaca sinica.* In R. H. Tuttle (ed.), *Socioecology and psychology of primates.* Mouton, The Hague. Pp. 125-151.

Dittus, W. P. J. 1977. The socioecological basis for the conservation of the toque monkey *(Macaca sinica)* of Sri Lanka. In Prince Rainier III. and G. H. Bourne (eds.), *Primate conservation.* Academic Press, New York. Pp. 238-265.

Dixson, A. F. 1981. *The Natural history of the gorilla.* Columbia University Press, New York.

Drickamer, L. C. and Vessey S. 1973. Group chaining in free-ranging male rhesus monkeys. *Primates* **14**: 359-368.

Drickamer, L. D. 1974. Social rank, observability, and sexual behaviour of rhesus monkeys *(Macaca mulatta). Journal of Reproduction and Fertility* **37**: 117-120.

Dunbar, R. I. M. 1979. Population demography, social organization, and mating strategies. In I. Bernstein and E. O. Smith (eds.), *Primate ecology and human origins.* Garland Press, New York. Pp. 65-88.

___________. 1983. Life history tactics and strategies of reproduction. In P. P. G. Bateson (ed.), *Mate choice.* Cambridge University Press, Cambridge. Pp. 423-433.

___________. 1984. *Reproductive decisions: An economic analysis of gelada baboon strategies.* Princeton University Press, New Jersey.

__________. 1986. The social ecology of gelada baboons. In R. W. Wrangham and D. Rubenstein (eds.), *Ecological aspects of social evolution.* Princeton University Press, Princeton. Pp. 322-335.

__________. 1988. *Primate social systems.* Comstock, Cornell University Press, Ithaca, New York.

Dunbar, R. I. M., and Dunbar, E. P. 1974. Ecological relations and niche seperation between sympatric terrestrial primates in Ethiopia. *Folia Primatologica* **21**: 36-60.

Eaton, G. G. 1974. Male dominance and agression in japanese macaque reproduction. In W. Montagna (ed.), *Reproductive behaviour.* Plenum Press, New York. Pp. 287-297.

Eglash, A. R. and Snowdon, C. T. 1983. Mirror-image responses in pygmy marmosets *(Cebuella pygmaea). American Journal of Primatology* **5**: 211-219.

Ehardt-Seward, C. and Bramblett, C. A. 1980. The Structure of social space among a captive group of vervet monkeys. *Folia Primatologica* **34**: 214-221.

Eibl-Eibesfeldt, I. 1974. *Ethology.* Holt, Rhinehart, Winston, New York.

__________. 1970. *Love and hate.* Holt, Rhinehart, Winston, New York.

Eisenberg, J. F. 1966. The social organization of mammals. *Handbook of Zoology* VIII **10**: 1-92.

__________. 1978. Comparative ecology and reproduction of new world monkeys. In D. G. Kleinman (ed.), *The Biology and conservation of the Callitrichidae.* Washington, Smithsonian Institution Press. Pp. 13-22.

Eisenberg, J. F., Muckenhirn, N. A. and Rudran, R. 1972. The relation between ecoogy and social structure in primates. *Science* **176**: 863-874.

Elizondo, R. 1977. Temperature regulation in primates. *International Review of Physiology and Environmental Physiology* **11**(5): 71-118.

Emlen, S. T. 1980. Ecological determinism and sociobiology. In G. Barlow and J. Silverberg (eds.), *Sociobiology: Beyond nature/nurture.* AAAS selected symposium. Pp. 125-140.

Emlen, S. T., and Oring, L. W. 1977. Ecology, sexual selection, and the evolution of mating systems. *Science* **197**: 215-223.

Enomoto, T. 1974. The sexual behaviour of Japanese monkeys *(Macaca fuscata). Journal of Human Evolution* **3**: 351-372.

__________. 1978. On social preference in sexual behaviour of Japanese monkeys *(Macaca fuscata). Journal of Human Evolution* **7**: 283-293.

Epple, G. 1978. Reproductive and social behaviour of marmosets with special reference to captive breeding. In F. Reinhardt and N. Gengozian (eds.), *Marmosets in experimental medicine.* Karger Press, Basel. Pp. 50-62.

Epple, G., Alveario, M. C. and Katz, Y. 1982. The Role of chemical communication in aggressive behavior and its gonadal control in the Tamarin *(Saquinis fuscicollis).* In C. T. Snowdon, C. H. Brown, and M. R. Peterson (eds.), *Primate communication.* Cambridge University Press, Cambridge. Pp. 259-268.

Ewer, R. F. 1960. Natural selection and neoteny. *Acta Biotheoretica* **13**: 161-184.

Fagen, R. 1981. *Animal play behavior.* Oxford University Press, New York.

Fairbanks, L. A. 1980. Relationships among adult females in captive vervet monkeys: Testing a model of rank-related attractiveness. *Animal Behaviour* **28**(3): 853-859.

Falk, D. 1987. Hominid paleoneurology. *Annual Review of Anthropology* **16**: 13-30.

Fedigan, L. and Fedigan, C. 1977. The Social development of a handicapped infant in a free living troop of japanese monkeys. In S. Chevalier-Skolnikoff and F. E. Poirier (eds.), *Primate biosocial development: Biological social and ecological determinants.* Garland Publishing Co, New York. Pp. 205-224.

__________. and Gouzoules, H. 1978. The consort relationships in a troop of japanese monkeys. In D. Chivers (ed.), *Recent advances in primatology.* London, Academic Press. Pp. 493-495.

__________. 1982. *Primate paradigms: Sex roles and social bonds.* Eden Press, Montreal.

__________. 1983. Dominance and reproductive success in primates. *Yearbook of Physical Anthropology* **26**: 91-129.

Feyerabend, P. 1975. *Against method: Outline of an anarchistic theory of knowledge.* Verso Press, London.

Fossey, D. 1983. *Gorillas in the Mist.* Houghton Mifflin, Boston.

Fouts, R. S. 1974. Language: Origins, definitions and chimpanzees. *Journal of Human Evolution* **3**:475-482.

Frisch, J. E. and S. J. 1968. Individual behavior and intertroop variability in japanese macaques. In P. Jay (ed.), *Primates: Studies in adaptation and variability.* Holt, Rinehart, Winston, New York. Pp. 243-252.

Furuya, Y. 1969. On the fission of troops of japanese monkeys. II. General view of troop fission of japanese monkeys. *Primates* **10**: 47-69.

Galdikas, B. M. F. 1978a. *Orangutan adaptation at Tanjung Puting Reserve, Central Borneo.* Unpublished doctoral dissertation. University of California, Los Angeles.

___________. 1978b. Orangutans and hominid evolution. In S. Udin (ed.), *Spectrum: Essays presented to Sutan Takdir Alisjahbana on his seventieth birthday.* Dian Rakyat, Jakarta.

___________. 1979. Orangutan adaptation at Tanjung Puting Reserve: Mating and ecology In D. A. Hamburg and E. R. McCown (eds.), *The great apes.* Benjamin/Cummings, Menlo Park. Pp. 194-233.

___________. 1981a. On pongid subsistence activites: Replies. *Current Anthropology* **22**(3): 316-320.

___________. 1981b. Orangutan reproduction in the Wild. In Graham, C. E. (ed.), *Reproductive biology of the great apes.* Academic Press, New York. Pp. 281-300.

___________. 1982. Orangutan tool use at Tanjung Puting Reserve, Central Indonesian Borneo (Kalimantan Tengah). *Journal of Human Evolution* **11**: 19-33.

___________. 1983. The orangutan long call and snag crashing at Tanjung Puting Reserve. *Primates* **24**: 317-384.

___________. 1984. Orangutan female sociality at Tanjung Puting. In: M. Small (ed.), *Female primates: Studies by women primatologists.* Alan R. Liss, New York. Pp. 217-235.

___________. 1985a. Orangutan sociality at Tanjung Puting Reserve, Kalimantan Tengah. *American Journal of Primatology* **9**: 101-119.

___________. 1985b. Adult male sociality and reproductive tactics among orangutans at Tanjung Puting. *Folia Primatologica* **45**: 9-24.

___________. 1985c. Subadult male orangutan sociality and reproductive behaviour. *American Journal of Primatology* 8: 87-99.

___________. 1988. Orangutan diet, range, and activity at Tanjung Puting, Central Borneo. *Journal of Primatology* **9**(1): 1-35.

___________. 1990. My life with orangutans. *International Wildlife* **20**(2): 34-41.

___________. and Teleki, G. 1981. Variations in subsistence activites of female and male pongids: New perspectives on the origins of hominid labor division. *Current Anthropology* **22**(3): 241-256.

___________. and Wood, J. 1990. Great ape and human birth intervals. *American Journal of Physical Anthropology* **83**(3): 185-192.

___________. and Insley, S. J. 1988. The fast call of the adult male orangutan. *Journal of Mammalogy* **69**(2): 371-375.

Galef, B. G. Jr. 1976. Social transmission of acquired behavior: A Discussion of tradition and social learning in vertebrates. In J. S. Rosenblatt, R. A. Hinde, E. Shaw and C. G. Beer (eds.), *Advances in the Study of Behavior.* **Vol**. 6. Academic Press, New York. Pp. 77-99.

Gallup, G. G. 1970. Chimpanzees: Self-recognition. *Science* **167**: 86-87.

___________. 1977. Self-recognition in primates, a comparative approach to the bidirectional properties of consciousness. *American Psychologist* **32**: 329-338.

___________. 1982. Self-awareness and the emergence of mind in primates. *American Journal of Primatology* **2**: 237-48.

Gartlan, J. S. 1968. Structure and function in primate society. *Folia Primatologica* **8**: 89-120.

Gartlan, J. S. and Struhsaker, T. T. 1972. Polyspecific associations and niche seperation of rain-forest anthropoids in Cameroon, West Africa. *West Africa Journal of Zoology, London* **168**: 221-266.

Gautier-Hion, A., Quris, R. and Gautier, J. P. 1983. Monospecific vs. polyspecific life: A comparative study of foraging and antipredatory tactics in a community of Cercopithecus monkeys. *Behavioral Ecology and Sociobiology* **12**: 325-335.

Geist, V. 1972. An Ecological and behavioural explanation of mammalian characteristics, and their implication to therapsid evolution. *Zeitschrift fr Sugetierkunde* **37**: 1-15.

___________. 1987. Bergmann's rule is invalid. *Canadian Journal of Zoology* **65**: 1035-1038.

Gibson, K. R. 1986. Cognition, brain size, and the extraction of embedded food resources. In J. G. Else and P. C. Lee (eds.), *Primate ontogeny, cognition and social behaviour.* Cambridge University Press, Cambridge. Pp. 93-103.

Gillan, D. J. 1981. Reasoning in the chimpanzee. II. Transitive inference. *Journal of Experimental Psychology and Animal Behaviour. Proceedings* **7**: 150-164.

Goodall, J. (van Lawick). 1968. A Preliminary report on expressive movements and communication in the Gombe Stream chimpanzees. In P. C. Jay (ed.), *Primate studies in adaptation and variability.* Holt, Rhinehart, Winston. Pp. 313-374.

__________. 1971. *In the Shadow of Man.* Houghton Mifflin, Boston.

__________. 1986. *The Chimpanzees of Gombe: Patterns of behaviour.* Harvard University Press, Cambridge, Mass.

Goodall, J. 1963. My life among wild chimpanzees. *National Geographic* **124**(2): 272-308.

__________. 1964. Tool use and aimed throwing in a community of chimpanzees. *Nature* **201**(4626): 1264-1266.

__________. Bandoro, A., Bergmann, E., Busse, C., Matama, H., Mpongo, E., Pierce, A. and Riss, D. 1979. Intercommunity interactions in the chimpanzee population of the Gombe National Park. In D. A. Hamburg and E. R. McCown (eds.), *The Great apes.* Benjamin/Cummings, Menlo Park California. Pp. 13-54.

Goss-Custard, J. D., Dunbar, R. I. M. and Aldrich-Blake, F. G. 1972. Survival, mating and rearing strategies in the evolution of primate social structure. *Folia Primatologica* **17**: 1-19.

Gould, S. J. 1971. Geometric similarity in allometric growth: A contribution to the problem of scaling in the evolution of size. *American Naturalist* **105**: 113-136.

__________. 1977. *Ontogeny and phylogeny.* The Belknap Press of Harvard University Press, Cambridge, Mass.

__________. 1980. Natural selection and the human brain. In *The panda's thumb.* W. W. Norton & Co., New York. Pp. 47-58.

__________. 1982. Darwinism and the expansion of evolutionary theory. *Science* **216**: 380-387.

Gould, S. J., and Lewontin, R. 1979. The spandrels of San Marco and the Panglossian paradigm: a critique of the adaptionist programme. Proceedings Royal Society of London. B. *Biological Science* **205**: 581-598.

Gouzoules, S., Gouzoules, H., and Marler, P. 1984. Rhesus monkey *(Macaca mulatta)* screams: Representational signalling in the recruitment of agonistic aid. *Animal Behavior* **32**: 182-193.

Green, S. 1975. Variations of vocal pattern with social situation in the Japanese monkey *(Macaca fuscata).* In L. A. Rosenblum (ed.), *Primate Behavior* Vol. 4. Academic Press, New York. Pp. 1-102.

__________. and Marler, P. M. 1979. The Analysis of animal communication. In P. Marler and J. G. Vandenbergh (eds.), *Handbook of behavioral neurobiology.* **Vol.** *3: Social behavior and communication.* Plenum Press, New York. Pp. 73-158.

Grewal, B. S. 1980. Social relationships between adult central males and kinship groups of Japanese monkeys at Arashiyama with some aspects of troop organization. *Primates* **21**: 161-180.

Griffin, D. 1976. *The Question of animal awareness: Evolutionary continuity of mental experience.* The Rockefeller University Press, New York.

__________. 1981. Prospects for a cognitive ethology. *Behavior and Brain Science* **1**: 527-538.

__________. 1984. *Animal thinking.* Harvard University Press, Cambridge.

Hall, K. R. L. 1963. Tool using performances as indications of behavioral adaptatbility. *Current Anthropology* **4**: 479-494.

__________. 1965. Social organization of the old world monkeys and apes. *Symposium of Zoological Society of London* **14**: 265-289.

__________. and Devore, I. 1965. Baboon social behaviour. In I. Devore (ed.), *Primate behaviour: Field studies of monkeys and apes.* Holt, Rinehart, Winston, New York. Pp. 53-110.

__________. 1968. *Social organization of hamadryas baboons: A Field study.* University of Chicago Press, Chicago.

Hamilton, W. D. 1963. The evolution of altruistic behaviour. *American Naturalist* **97**: 354-356.

__________. 1964. The genetical theory of social behaviour. *Journal of Theoretical Biology* **7**: 1-52.

__________. 1971. Geometry of the selfish herd. *Journal of Theoretical Biology* **31**: 295-311.

__________. 1972. *Life's color code.* Prentice Hall, New Jersey.

Hamilton, W. J., Buskirk, R. E. and Buskirk, W. H. 1976. Defence of space and resources by chacma *(Papio ursinus)* baboons in an African desert and swamp. *Ecology* **57**: 12 64-72.

Hanby, J. P., Robertson, L. T. and Phoenix, C. H. 1971. The sexual behaviour of a confined troop of japanese macaques. *Folia Primatologica* **16**: 123-143.

Harcourt, A. H. 1978. Activity periods and patterns of social interactions: A Neglected problem. *Behaviour* **66**(1/2): 121-135.

___________. 1979. Contrasts between male relationships in wild gorilla groups. *Behavioural Ecology and Sociobiology* **5**: 39-49.

___________. 1988. Alliances in contest and social intelligence. In R. Byrne and A. Whiten (eds.), *Machiavellian intelligence: Social expertise and the evolution of intellect in monkeys, apes and humans.* Oxford University Press, Oxford. Pp. 132-152.

___________. and K. J. Stewart. 1989. Functions of alliances in contests within wild gorilla groups. *Behaviour* **109**(3/4): 176-190.

Harré R. and Secord, P. F. 1972. *The Explanation of social behaviour.* Blackwell, Oxford.

Harris, M. 1987. *Cultural anthropology* [2nd ed.]. Harper & Row, New York.

Harrison, G. A., Tanner, J. M., Pilbeam, D. R. and Baker, P. T. 1988. *Human biology: An Introduction to human evolution, variation, growth, and adaptability* [3rd ed.]. Oxford University Press, Oxford.

Hauser, M. D. 1988. Invention and social transmission: New data from wild vervet monkeys. In R. Byrne and A. Whiten (eds.), *Machiavellian intelligence: social expertise and the evolution of intellect in monkeys, apes and humans.* Clarendon Press, Oxford. Pp. 327-343.

Hausfater, G. 1975. Dominance and reproduction in baboons *(Papio cynocephalus). Contributions to Primatology* **7**: 1-35.

___________. 1976. Predatory behaviour of yellow baboons. *Behaviour* **54**: 1-2.

___________. Altmann, J. and Altmann, S. A. 1982. Long-term consistency of dominance relations among female Baboons. *Science* **217**: 752-755.

___________. and Takacs, D. 1987. Structure and function of hindquarter presentations in yellow baboons *(Papio cynocephalus). Ethology* **74**(4): 297-319.

Hays, W. 1973. *Statistics for the social sciences.* [2nd ed.]. Holt, Rinehart, Winston, New York.

Headrick, M. W., Feather, B. W. and Wells, D. T. 1970. Voluntary changes from baseline heart rate with augmented sensory feedback. *Psychophysiology* **6**: 636.

Hendrichs, H. 1983. On the evolution of social structure in mammals. In J. F. Eisenberg and D. G. Kleinman (eds.), *Advances in the study of mammalian behaviour.* The American Society of Mammalogists, Pittsburgh. Pp. 738-750.

Hill, R. W. 1975. *Comparative physiology of animals: An Environmental approach.* Harper and Row, New York.

Hinde, R. A. and Spencer-Booth, Y. 1967. The Effect of social companions on mother-infant relations in rhesus monkeys. In D. Morris (ed.), *Primate ethology.* Weidenfeld and Nicolson, London. Pp. 267-286.

__________. and Atkinson, S. 1970. Assessing the roles of social partners in maintaining mutual proximity, as exemplified by mother-infant relations in rhesus monkeys. *Animal Behavior* **18**: 169-176.

__________. and Steveson-Hinde, J. 1976. Towards understanding relationships: dynamic stability. In P. P. G. Bateson and R. A. Hinde (eds.), *Growing points in ethology.* Cambridge: Cambridge University Press. Pp. 451-479.

__________. (ed.). 1983. *Primate social relationships: An Integrated approach.* Blackwell Scientific, Oxford. Pp. 209-219.

__________. 1983b. Matriline membership and male rhesus reaching high ranks in their natal troops. In R. A. Hinde (ed.), *Primate social relationships: An integrated approach.* Blackwell, Oxford. Pp. 171-175.

Hiraiwa-Hasegawa, M. 1990. Role of food sharing between mother and infant in the ontogeny of feeding behavior. In T. Nishida (ed.), *The chimpanzees of the Mahale Mountains: Sexual and life History Strategies.* Univerisity of Tokyo Press, Toyko. Pp. 267-275.

Hladik, C. M. 1975. Ecology, diet and social patterning in old and new world primates. In R. H. Tuttle (ed.), *Socioecology and psychology of primates.* Mouton, The Hague. Pp. 3-35.

Homans, G. C. 1961. *Social behaviour: Its elementary forms.* Routledge and Kegan Paul, London.

Hooff, J. A. R. A. M. 1962. Facial expressions in higher primates. *Symposium of the Zoological Society of London* **8**: 97-125.

Hoogland, J. L. 1979. The effects of colony size on individual alertness of prairie dogs *(Sciuridae: Crynomys* spp). *Animal Behaviour* **27**: 394-407.

Hori, T., Tokura, H. and Tadaki, E. 1972. Surface area in the Japanese Monkey, *(Macaca fuscata). Journal of Applied Physiology* **32**(3): 409-411.

Hornshaw, S. G. 1975. *Animal proxemics: A Study of the use of space in Macaca silenus at the Assiniboine Park Zoo.* University of Manitoba Anthropology Papers, Manitoba.

___________. 1982. A comparative analysis of socio-spatial behaviour in a captive and a non-captive group of Macaca sylvanus. *Ph.D. Dissertation.* University of Toronto, Toronto.

___________. 1985a. Proximity behavior in a captive group of Lion-tailed macaques *(Macaca silenus).* In P. G. Heltne (ed.), *The lion-tailed Macaque: status and conservation.* Alan R. Liss, Inc, New York. Pp. 269-292.

___________. 1985b. A Comparison of proximity behaviour in two groups of barbary macaques. In J. E. Fa (ed.), Plenum Press, New York. Pp. 221-239.

Horr, D. A. 1977. Orangutan maturation: Growing up in a female world. In S. Chevalier-Skolnikoff and R. E. Poirier (eds.), *Primate bio-social development: Biological, social and ecological determinants.* Garland, New York. Pp. 289-321.

___________. 1975. The Borneo orangutan: Population structure and dynamics in relationship to ecology and reproductive strategy. In L. A. Rosenblum (ed.), *Primate behavior: Developments in field and laboratory research.* **Vol.** 4. Academic Press, New York. Pp. 25-80.

Hrdy, S. B. 1977. *The Langurs of* Abu: *Male and female strategies of reproduction.* Harvard University Press, Cambridge.

___________. and W. Bennett. 1979. The fig connection. *Harvard Magazine* **09/10**: 25-30.

___________. 1984. Comparative and evolutionary perspectives on infanticide: An Introduction and overview. In G. Hausfater and S. B. Hrdy (eds.), *Infanticide: Comparative and evolutionary perspectives.* Aldine, Hawthorne, New York.

___________. and Whiten, P. L. 1987. Patterning of sexual activity. In B. Smuts, D. L. Cheney, R. M. Seyfarth, P. W. Wrangham and T. T. Struhsaker (eds.), *Primate societies.* University of Chicago Press, Chicago. Pp. 370-384.

Humphrey, N. K. 1976. The social function of intellect. In P. G. Bateson and R. A. Hinde (eds.), *Growing points in ethology.* Cambridge University Press, Cambridge. Pp. 303-317.

Huxley, J. S. 1932. *Problems of relative growth.* Methuen, London.

Irving, L. 1957. The Usefulness of Scholander's views on adaptive insulation of mammals. *Evolution* **11**: 257-260.

___________. 1972. *Artic life of birds and mammals*. Springer-Verlag, New York.

Isaac, G. 1978. The food-sharing behaviour of protohuman hominids. *Scientific American* **238**(4): 90-108.

Itani, J. 1959. Parental care in the wild japanese monkey, *Macaca fuscata. Primates* **2**: 61-93.

___________. 1965. On the acquisition and propagation of a new food habit in the troop of japanese monkeys at Takasakiyama. In K. Imanishi and S. A. Altmann (eds.), *Japanese monkeys.* Emory University Press, Atlanta. Pp. 52-65.

___________. and Nishimura, A. 1973. The study of infrahuman culture in Japan. In E. W. Menzel, Jr. (ed.), *Precultural primate behavior.* Symposium of the 4th Cong. International Society of Primates. S. Karger, Basel. **1**: 26-50.

Itoigawa, N. 1974. Variables in male leaving a group of Japanese monkeys. In S. Kondo (ed.), *Proceedings Symposium of the 5th Cong. of the International Primatological Society.* Japan Science Press, Tokyo. Pp. 233-245.

Jamieson, I. G. 1986. The Functional approach to behaviour: Is it useful? *American Naturalist.* **127**: 195-208.

Jerison, H. 1983. The evolution of the mammalian brain as an information processing system. In J. F. Eisenberg and D. G. Kleiman (eds.), *Advances in the study of mammalian behaviour.* The American Society of Mammalogists, Pittsburgh. Pp. 113-146.

___________. 1985. Animal intelligence as encephalisation. In L. Weiskrantz (ed.), *Animal intelligence.* Clarendon Press, Oxford. Pp. 21-35.

John, B. and Miklos, G. L. G. 1988. *The Eukaryote genome in development and evolution.* Allen and Unwin, London.

Jolly, A. 1966. Lemur social behavior and primate intelligence. *Science* **153**: 501-506.

___________. 1985. *The Evolution of primate behavior.* [2nd ed.]. Macmillan, Toronto.

___________. 1985. The evolution of primate behaviour. *American Scientist* **73**: 230-239.

Jones, E. E. and Gerard, H. B. 1967. *Foundations of social psychology.* Wiley, New York.

Jorde, L. B. and Spuhler, J. N. 1974. A statistical analysis of selected aspects of primate demography, ecology, and social behaviour. *Journal of Anthropological Research* **30**: 149-224.

Joreskog, K. C., Sorbom, D. and Magidam, J. 1979. *Advances in factor analysis and structural equation models.* Abt books, Cambridge.

Kano, T. 1982. The use of leafy twigs for rain cover by the pygmy chimpanzees of Wamba. *Primates* **23**(3): 453-457.

Kaplan, J. R. 1978. Fight interference and altruism in rhesus monkeys. *American Journal of Physical Anthropology* **49**: 241-250.

Kawai, M. 1965. Newly acquired precultural behaviour of the natural troop of japanese monkeys on Koshima Islet. *Primates* **6**: 1-30.

Kleiber, M. 1961. *The Fire of life: An Introduction to animal energetics.* Wiley and Sons, New York.

Kleinman, D. G. 1977. Monogamy in mammals. *Quarterly Review of Biology* **52**: 208-224.

Kortlandt, A. 1967. Experimentation with chimpanzees in the wild. In D. Starck, R.Schneider, and H.J. Kuhn. (eds.), *Progress in Primatology.* Fisher, Stuttgart. Pp. 208-224.

Koyama, N. 1970. Changes in dominance rank and division of a wild japanese monkey troop in Arashiyama. *Primates* **11**: 325-390.

Kummer, H. 1967. Tripartite relations in hamadryas baboons. In Altmann, S. (ed.), *Social communication among primates.* University of Chicago Press, Chicago. Pp. 63-72.

____________. 1968. *Social organization of hamadryas baboons: A Field study.* University of Chicago Press, Chicago.

____________. 1971. *Primate Societies: Group techniques of ecological adaptation.* Aldine, Chicago.

____________. 1978. On the value of social relationships to non-human primates: A Heuristic scheme. *Social Science Information* **17**(4/5): 687-705.

____________. 1982. Social knowledge in free-ranging primates. In D. R. Griffin (ed.), *Animal mind - Human mind.* Springer-Verlag, Berlin. Pp. 113-130.

____________. 1968. Two variations in the social organization of baboons. In P. Jay (ed.), *Primates: Studies in adaptation and variability.* Holt, Rinehart, Winston, San Francisco. Pp. 293-312.

___________. Dasser, V. and Hoyningen-Huene, P. 1990. Exploring primate social cognition: Some critical remarks. *Behaviour* **112**(1-2): 84-97.

Kurland, J. A. 1977. Kin selection in the japanese macaque. *Contributions to Primatology* **12**. S. Karger, Basel.

___________. and Gaulin, S. J. C. 1984. The Evolution of male parental investment: Effects of genetic relatedness and feeding ecology on the allocation of reproductive effort. In D. M. Taub (ed.), *Primate paternalism.* Van Nostrand Reinhold, New York.

Kuroda, S. 1984. Interactions over food among pygmy chimpanzees. In R. L. Susman (ed.), *The Pygmy chimpanzee: Evolutionary biology and behaviour.* Plenum Press, New York. Pp. 301-324.

Lack, D. 1968. *Ecological adaptations for breeding in birds.* Methuen, London.

Lehrman, D. S. 1970. Semantic and conceptual issues in the nature-nurture Problem. In L. R. Aronson, E. Tobach, D. S. Lehrman, and J. S. Rosenblatt (eds.), *Development and evolution of behaviour.* Freeman, San Francisco. Pp. 17-52.

Lethmate, J. 1982. Tool-using skills of orangutans. *Journal of Human Evolution* **11**: 49-64.

Lewontin, R. C. 1974. *The Genetic basis of evolutionary change.* Columbia University Press, New York.

Lindburg, D. G. 1977. Feeding behaviour and diet of rhesus monkeys *(Macaca mulatta)* in a Siwalik forest in North India. In T. H. Clutton-Brock (ed.), *Primate ecology.* Academic Press, New York. Pp. 223-249.

Lipetz, V. E., and Bekoff, M. 1982. Group size and vigilance in pronghorns. *Zoologische Tierpsychologie* **58**: 203-216.

Lloyd Morgan, C. 1896. *Habit and instinct.* Edward Arnold, London.

Loewenstein, L. and Zihlmann, A. 1988. The Invisible ape. *New Scientist* **120**: 56-59.

Lorenz, K. Z. 1965. *Evolution and modification of behavior.* University of Chicago Press, Chicago.

___________. 1978. *The Ape within us.* Collins, London.

MacKinnon, J. 1974. The Behaviour and ecology of wild orangutans *(Pongo pygmaeus). Animal Behaviour* **22**: 2-74.

Manley, G. H. 1974. Functions of the external genital glands of *Perodicticus* and *Arctocebus*. In R. D. Martin, G. A. Doyle and A. C. Walker (eds.), *Prosimian biology*. Duckworth, London. Pp. 313-330.

Marler, P. M. and Hobbett, L. 1975. Individuality in long range vocalization of wild chimpanzees. *Zeitschrift fur Tierpsychologie* **38**: 97-109.

Maruhashi, T. 1982. An ecological study of troop fissions of Japanese monkeys *(Macaca fuscata yakui)* on Yakushima Island, Japan. *Primates* **23**: 317-337.

Maruyama, M. 1978. Prigogine's epistemology and its implications for the social sciences. *Current Anthropology* **19**(2): 453-454.

Mason, W. A. 1978. Ontogeny of social systems. In D. Chivers (ed.), *Recent Advances in Primatology*. Academic Press, New York. Pp. 5-14.

___________. 1982. Primate social intelligence: contributions from the laboratory. In D. Griffin (ed.), *Animal Mind - Human Mind*. Springer-Verlag, Berlin. Pp. 131-140.

___________. 1986. Can primate political traits be identified? In Else, J. G. and P. C. Lee (eds.), *Primate ontogeny cognition and social behaviour*. Cambridge University Press, Cambridge. Pp. 3-12.

Mauss, M. 1925. *Essai sur le Don*. (English translation [1954]: *The gift*. By I. Cunison.) The Free Press, New York.

Maynard-Smith, J. 1977. Parental investment - a prospective analysis. *Animal Behaviour* **25**: 1-9.

Mayr, E. 1956. Geographical character gradients and climatic adaptation. *Evolution* **10**: 823-824.

___________. 1960. The Emergence of evolutionary novelties. In S. Tax (ed.), *Evolution after Darwin. 1: The Evolution of life*. University of Chicago Press, Chicago. Pp. 349-380.

___________. 1963. *Animal species and evolution*. Belknap, Harvard University Press, Cambridge.

___________. 1970. *Populations, species and evolution*. Belknap, Harvard University Press, Cambridge.

___________. 1972. Sexual selection and natural selection. In B. G. Campbell (ed.), *Sexual selection and the descent of man 1871-1971*. Aldine Press, Chicago. Pp. 87-104.

___________. and Provine, W. B. 1980. *The Evolutionary synthesis*. Harvard University Press, Cambridge, Mass.

McFarland, D. 1987. *The Oxford companion to animal behavior.* Oxford, Oxford University Press.

McFeat, T. 1974. *Small group cultures.* Pergamon Press, New York.

McGrew, W. C. 1979. Evolutionary implication of sex differences in chimpanzee predation and tool use. In Hamburg, D. A. and E. R. McCown (eds.), *The Great apes: Perspectives on human evolution.* **Vol.** 5. Benjamin Cummings, Menlo Park. Pp. 441-463.

___________. 1987. Tools to get food: The Subsistence of Tasmanian aborigines and Tanzanian chimpanzees compared. *Journal of Anthropological Research* **43**(3): 247-258.

___________. 1989. Why is ape tool use so confusing? In Standen, V. and R. A. Foley (eds.), *Comparative socioecology.* Blackwell Scientific, Oxford. Pp. 457-472.

___________ and Tutin, C. E. G. 1978. Evidence for a social custom in wild chimpanzees? *Man* **13**: 234-251.

___________. and Tutin, C. E. G. and Baldwin, P. J. 1979. Chimpanzees, tools, and termites: Cross-cultural comparisons of Senegal, Tanzania, and Rio Muni. *Man* **14**: 185-214.

McNab, B. K. 1970. Body weight energetics of temperature regulation. *Experimental Biology* **53**: 329-348.

___________. 1971. Communication about the environment in a group of young chimpanzees. *Folia Primatologia* **15**: 220-232.

___________. 1971. On the ecological significance of Bergmann's Rule. *Ecology* **52**(5): 845-854.

Meikle, D. B., and Vessey, S. H. 1981. Nepotism among rhesus monkey brothers. Nature **294**: 160-161.

Menard, N., Hecham, R., Vallet, D. and Chikhi, H. 1990. Grouping patterns of a mountain population of *M. sylvanus* in Algeria – A fission-fusion system? *Folia Primatologica* **55**: 166-175.

Menzel, C. R. 1991. Cognitive aspects of foraging in japanese monkeys. *Animal Behaviour* **41**: 397-402.

Menzel, E. W. Jr. 1974. A Group of chimpanzees in a one-acre field. In Schrier, A. M., H. F. Harlow, and F. Stollnitz (eds.), *Behaviour of non-human primates.* Academic Press, New York. Pp.83-153.

___________. 1978. Cognitive mapping in chimpanzees. In S. H. Hulse, H. Fowler and W. K. Honig (eds.), *Cognitive process in animal behavior.*

behavior. Lawrence Erlbaum Assoc., Hillsdale, New Jersey. Pp. 375-422.

___________. Premack, D. and Woodruff, G. 1978. Map reading by chimpanzees. *Folia Primatologia* **19**: 241-249.

___________. 1979. Communication of object locations in a group of young chimpanzees. In D. A. Hamburg and E. R. McCown. *The Great apes.* Benjamin Cummings, Menlo Park California. Pp. 358-371.

___________. and Wyers, E. J. 1981. Cognitive aspects of foraging behavior. In A. C. Kamil and T. D. Sargent (eds.), *Foraging behavior.* Garland, New York. Pp. 355-377.

___________. and Juno, C. 1985. Social foraging in marmoset monkeys and the question of intelligence. In L. Weiskrantz (ed.), *Animal intelligence.* Clarendon Press, Oxford. Pp. 145-158.

___________. Savage-Rumbaugh, E. S. and Lawson, J. 1985. Chimpanzee *(Pan troglodytes)* spatial problem solving with the use of mirrors and televised equivalents of mirrors. *Journal of Comp. Psychology* **99**: 211-217.

Miles, L. H. W. 1990. The Cognitive foundations for reference in a signing orangutan. In S. T. Parker and K. R. Gibson (eds.), *Language and* intelligence in *monkeys and apes: Comparative developmental perspectives.* Cambridge University Press, Cambridge. Pp. 511-539.

Miller, N. E. 1969. Learning of visceral and glandular responses. *Science* **163**: 434-445.

Miller, R. E. 1967. Experimental approaches to the physiological and behavioral concomitants of affective communication in the rhesus monkey. In S. A. Altmann (ed.), *Social communication among primates.* University of Chicago Press, Chicago. Pp. 125-134.

___________. 1972. Experimental studies of communication in the monkey. In L. A. Rosenblum (ed.), *Primate behavior: Developments in field and laboratory research.* Academic Press, New York. Pp. 139-175.

___________. 1975. Non-verbal expression of aggression and submission in social groups of primates. In P. Pliner, T. Kromes, and T. Alloway (eds.), *Non-verbal communication of aggression.* Plenum Press, New York. Pp. 135-160.

Milton, K. 1981. Distribution patterns of tropical plant foods as an evolutionary stimulus to primate mental development. *American Anthropologist* **83**: 534-548.

___________. 1988. Foraging behaviour and the evolution of primate intelligence. In R. Byrne and A. Whiten (eds.), *Machiavellian intelligence: Social expertise and the evolution of intellect in monkeys, apes and humans.* Oxford University Press, Oxford. Pp. 285-305.

Missakian, E. A. 1973. Genealogical and cross-genealogical dominance relations in a group of free-ranging rhesus monkeys *(Macaca mulatta)* on Cayo Santiago. *Primates* **13**: 169-180.

Mitani, J. C. 1985a. Mating behaviour of male orangutans in the Kutai Reserve, East Kalimantan, Indonesia. *Animal Behaviour* **33**: 392-402.

___________. 1985b. Sexual selection and adult male orangutan long calls. *Animal Behaviour* **33**: 272-283.

___________. 1990. Experimental field studies of asian ape social systems. *International Journal of Primatology* **11**(2): 103-126.

Miyadi, D. 1964. Social life of japanese monkeys. *Science* **143**: 783-786.

Muckenhirn, N. A. and Eisenberg, J. F. 1973. Home ranges and predation of the Ceylon leopard. In R. L. Eaton (ed.), *World cats.* World Wildlife Safari, Winston, Oregon. Pp. 142-175.

Mller, E. F. and Jaksche, H. 1980. Thermoregulation, oxygen consumption, heart rate and evaporative water loss in the thick-tailed bushbaby *(Galago crassicaudatus). Zeitshrift fr Sugetierkunde* **45**(5): 269-278.

Nagel, U. 1971. Social organization in baboon hybid zone. *Proceedings of the 3rd International Congress of Primatology, Zurich* **3**: 48-57.

___________. 1973. A Comparison of anubis baboons, hamadryas baboons and their hybrids at a species border in Ethiopia. *Folia Primatologica* **19**: 104-165.

Nakamichi, M. and Yoshida, A. 1986. Discrimination of mother by infant among japanese macaques *(Macaca fuscata). International Journal of Primatology* **7**(5): 481-490.

Napier, J. 1970. *The Roots of mankind.* Smithsonian Press, Washington, D. C.

Nishida, T. 1980. The Leaf-clipping display: A Newly discovered expressive gesture in wild chimpanzees. *Journal of Human Evolution* **9**: 117-128.

___________. 1987. Local traditions and cultural transmission. In B. Smuts, D. L. Cheney, R. M. Seyfarth, P. W. Wrangham and T. T. Struhsaker (eds.), *Primate societies.* University of Chicago Press, Chicago. Pp. 462-474.

Okano, T. 1965. Preliminary survey of the orangutan in North Borneo. *Primates* **6**(1): 123-128.

Orians, G. H. 1969. On the evolution of mating systems in birds and mammals. *American Naturalist* **103**: 589-603.

Oring, L. W. 1982. Avian mating systems. In J. K. D. Farner and K. Parkers (eds.), *Avian biology.* Academic Press, New York.

Packer, C . 1979. Inter-troop transfer and inbreeding avoidance in *Papio anubis. Animal Behaviour* **27**: 1-36.

___________. 1984. Dispersal and Inbreeding Avoidance. *Animal Behaviour* **33**(2): 676-678.

___________. 1977. Reciprocal altruism in *Papio anubis*. *Nature* **265**: 441-443.

Parker, S. T. 1987. A Sexual selection model for hominid evolution. *Human Evolution* **2**: 235-253.

___________. and Gibson, K. R. 1979. A Developmental model for the evolution of language and intelligence in early hominids. *Brain Behaviour Science* **2**: 367-408.

Parker, C. E. 1978. Opportunism and the rise of intelligence. *Journal of Human Evolution* **7**: 596-608.

___________. 1977. Object manipulation, tool use and sensorimotor intelligence as feeding adaptation in Cebus monkeys and great apes. *Journal of Human Evolution* **6**: 623-641.

Passingham, R. 1982. *The Human primate.* W. H. Freeman & Co., Oxford.

Paterson, J. D. [In press.] Bergmann's Rule is Invalid - A Reply to Geist. *Canadian Journal of Zoology* **3**(2): 54-64.

___________. 1973. Ecologically differentiated patterns of aggressive and sexual behavior in two troops of Ugandan Baboons *(Papio anubis). American Journal of Physical Anthropology* **38**(2): 641-647.

___________. 1976. Variations in ecology and adaptation of Ugandan baboons *(Papio cynocephalus anubis):* With special reference to forest environments and analog models for early hominids. *Ph.D. Thesis.* University of Toronto, Toronto.

___________. 1980. A Preliminary study of the relationships between ambient environment and positional thermoregulatory behaviour in *Alouatta palliata. Antropologica Contemporanea.* **Vol.** 3(2): 254-259.

___________. 1982. Thermoregulatory behaviour and ecological factors in primates. *Canadian Review of Physical Anthropology* **3**: 3-11.

___________. 1986. Shape as a factor in primate thermoregulation. In D. M. Taub and F. A. King (eds.), *Current perspectives in primate social dynamics.* Van Nostrand, New York. Pp. 228-242.

___________. 1988. Computer based data collection. *Canadian Review of Physical Anthropology* **6**(2): 55-62.

Patterson, F. G. 1978. Conversations with a gorilla. *National Geographic* **154**: 438-465.

___________. and Linden, E. 1981. *The Education of Koko.* Holt, Rinehart, Winston, New York.

Peters, R. H. 1983. *The Ecological implications of body size.* Cambridge University Press, Cambridge.

Piaget, J. 1978. *Behavior and evolution.* Pantheon Books, New York.

Platt, M. M. and Thompson, R. L. 1985. Mirror responses in Japanese Macaque troop (Arashiyama West). *Primates* **26**(3): 300-314.

Poirier, F. E. 1972. Introduction. In Poirier, F. E. (ed.), *Primate socialization.* Random House, New York. Pp. 3-28.

Post, D., Hausfater, G. and McCuskey. 1980. Feeding behaviour of yellow baboons *(Papio cynocephalus):* relationships to age, gender, and dominance rank. *Folia Primatologica* **34**: 170-195.

Prigogine, I. and Stengers, I. 1984. *Order out of chaos: Man's new dialogue with nature.* Bantam Books, Toronto.

Quiatt, D. 1987. Household structure in an extended family of gibbons *(Hylobates lar).* Abstract, *American Journal of Primatology* (Presented at 10th Meeting of American Society of Primatologists, June 1987). P. 365.

___________. 1988. Regulation of mating choice in nonhuman primates. In C. G. N. Masci-Taylor and A. J. Boyce (eds.), *Human mating patterns.* Cambridge University Press, Cambridge. Pp. 120-140.

Ransom, T. W. 1981. *Beach troop of the Gombe.* Bucknell University Press, Lewisburg.

___________. and Rowell, T. E. 1972. Early social development of feral baboons. In Poirier, F. E. (ed.), *Primate socialization.* Random House, New York. Pp. 105-144.

Rasmussen, D. 1983. Evolutionary, proximate, and functional primate social ecology. In P. P. G. Bateson and P. Klopfer (eds.), *Perspectives in ethology*. Plenum Press, New York. Pp. 75-103.

Reinig, W. F. 1939. Besteht die Bergmannishe Regel zu Recht? *Arch. Naturgesch. N.F.* **8**: 70-88.

Reisenfeld, A. 1981. The Role of body mass in thermoregulation. *American Journal of Physical Anthropology* **55** (1): 95-100.

Reusch, J. and Bateson, G. 1968. Individual, group and culture: A Review of the theory of human communication. In J. Reusch and G. Bateson (eds.), *Communication: The Social matrix of psychiatry*. W. W. Norton and Company, Inc., New York. Pp. 273-289.

Reynolds, P. C. 1981. *On the evolution of human behaviour: The Argument from animals to man.* University of California Press, Berkeley.

Reynolds, V. 1968. *The Apes.* Cassel & Co., London.

___________. 1970. Roles and role change in monkey society: The Consort relationships of rhesus monkeys. *Man(N.S.)* **5**(3): 449-463.

___________. 1986. Primate social thinking. In J. G. Else and P. C. Lee (eds), *Primate ontogeny, cognition and social behaviour: The Argument form animals to man.* Cambridge University Press, Cambridge. Pp. 53-60.

___________. 1976. The origins of a behavioural vocabulary: the case of the rhesus monkey. *Journal of Theoretical Social Behaviour* **6**: 105-142.

Richard, A. J. 1985. *Primates in nature.* Freeman, New York.

___________. and Schulman, R. 1982. Sociobiology: Primate field studies. *Annual Review of Anthropology* **11**: 231-255.

___________. Goldstein, S. J. and Dewar, R. E. 1989. Weed macaques: the evolutionary implications of macaque feeding ecology. *International Journal of Primatology* **106**: 569-597.

Rijksen, H. J. 1978. *A Field study on Sumatran Orangutans (Pongo pygmaeus abelii) Lesson 1827.* Veeman & Zonen, Wageningen.

Robinson, J. G. 1981. Spatial structure in foraging groups of wedge-capped capuchin monkeys (*Cebus nigrivittatus*). *Animal Behaviour* **29**: 1036-1056.

Rodman, P. S. 1973. Population composition and adaptive organisation among orangutans of the Kutai Reserve. In R. Michael and J. Crook. *Comparative ecology and behaviour of primates.* Academic Press, London. Pp. 171-209.

___________. 1979. Individual activity patterns and the solitary nature of orangutans. In D. A. Hamburg and E. R. McCown. *The Great apes.* Benjamin/Cummings, Menlo Park. Pp. 234-255.

___________. and Mitani, J. C. 1987. Orangutans: Sexual dimorphism in a solitary species. In B. Smuts, D. L. Cheney, R. M. Seyfarth, R. W. Wrangham and T. T. Struhsaker (eds.), *Primate societies.* University of Chicago Press, Chicago. Pp. 146-155.

Roitblatt, H. L., Bever, T. G. and Terrace, H. S. (eds.), 1984. *Animal cognition.* Lawrence Erlbaum Associates, Hillsdale New Jersey.

Rondinelli, R. and Klein, L. L. 1976. An Analysis of adult spacing tendencies and related social interactions in a colony of spider monkeys *(Ateles geoffroyi)* at the San Francisco Zoo. *Folia Primatologica* **25-26**: 122-142.

Rowell, T. E. 1966. Forest-living baboons in Uganda. *Journal of Zoology, London* **149**: 344-364.

___________. 1966. Hierarchy in the organization of a captive baboon group. *Animal Behavior* **14**: 430-433.

___________. 1969. Variability in the social organization of primates. In D. Morris (ed.), *Primate ethology.* Anchor Books, New York. Pp. 283-305.

___________. 1972. *The Social behaviour of monkeys.* Penguin Books, Harmondsworth.

___________. 1987. On the significance of the concept of the harem when applied to animals. *Social Science Information* **26**: 649-669.

___________. 1988. Beyond the one-male group. *Behaviour* **104**: 191-201.

___________ and Olson, D. K. 1983. Alternative mechanism of social organization in monkeys. *Behaviour* **86**: 31-54.

___________. and Hartwell, K. M. 1978. The interaction of behaviour and reproductive cycles in patas monkeys. *Behavioral Biology* **24**: 141-167.

Rumbaugh, D. and Pate, J. L. 1984. The Evolution of cognition in primates. In H. L. Roitblatt, T. G. Bever and H. S. Terrace. *Animal cognition.* Lawrence Erlbaum Assoc. Hillsdale, New Jersey. Pp. 569-587.

Rumbaugh, D. M. 1986. Animal thinking - by stimulation or simulation? In J. G. Else and P. C. Lee (eds.), *Primate ontogeny, cognition and social behaviour.* Cambridge University Press, Cambridge. Pp. 31-42.

Saayman, G. S. 1971. Behaviour of the adult males in a troop of free ranging chacma baboons *(Papio ursinus). Folia Primatologica* **15**: 36-57.

Sacher, G. A. 1982. The Role of brain maturation in the evolution of the primates. In E. Armstrong and Dean Falk (eds.), *Primate brain evolution.* Plenum, New York. Pp. 97-112.

Sackett, G. 1966. Monkeys reared in isolation with pictures as visual input: Evidence for an innate releasing mechanism. *Science* **154**: 1470-1473.

___________. 1968. Abnormal behaviour in laboratory reared rhesus monkeys. In M. W. Fox (ed.), *Abnormal behavior in animals* W. B. Saunders Co., Philadelphia. Pp. 33-54.

Sade, D. S. 1965. Some aspects of parent-offspring and sibling relations in a group of rhesus monkeys, with a discussion of grooming. *American Journal of Physical Anthropology* **23**: 1-18.

___________. 1972. Sociometrics of *Macaca mulatta. 1.* Linkages and cliques in grooming matrices. *Folia Primatologica* **18**: 196-223.

___________. 1980. Can "fitness" be measured in primate populations? In H. Markl (ed.), *Evolution of social behaviour: Hypotheses and empirical tests.* Verlag Chemie, Basel. Pp. 97-114.

Sarrus et Rameaux. 1837. *Bulletin de l'academie de Medicine* **2**: 538.

Savage-Rumbaugh, E. S., Romski, M.A. Hopkins, W. D. and Sevcik, R. A. 1989. Symbol acquisition and use by *Pan troglodytes, Pan paniscus, Homo sapiens.* In P. G. Heltne and L. A. Marquardt (eds.), *Understanding chimpanzees.* Chicago Academy of Sciences, special publication no. 16. Harvard University Press, Harvard. Pp. 266-295.

Savage-Rumbaugh, S., Scanlon, J. L. and Rumbaugh, D. 1980. Communicative intentionality in the chimpanzee. *Brain and Behavioural Sciences* **30**: 620-623.

Sawaguchi, T. 1990. Relative brain size, stratification, and social structure in anthropoids. *Primates* **31**(2): 257-272.

___________. 1965. The Behaviour of the Mountain Gorilla. In I. Devore (ed.), *Primate Behavior - Field Studies of Monkeys and Apes.* Rinehart and Winston, New York. Pp. 324-367.

Schaller, G. B. 1961. The orangutan in Sarawak. *Zoologica* **46**(2): 73-82.

Schmidt-Nielsen, K. 1970. Energy metabolism, body size, and problems of scaling. *Federation Proceedings* **29**: 1524-1532.

___________. 1975. Scaling in biology: The Consequences of *size*. *Journal of Experimental Biology* **194**(1): 287-307.

___________. 1984. *Scaling: Why is animal size so important?* Cambridge University Press, Cambridge.

Schneirla, T. C. 1955. Evolution of climatic adaptation in homeotherms. *Evolution* **9**: 15-26.

___________. 1956. The Interrelationships of the "innate" and the "acquired" in instinctive behaviour. In P. P. Grass (ed.), *L'Instinct dans le comportement des animaus et de l'homme.* Masson, Paris.

___________. 1957. The concept of development in comparative psychology. In D. B. Harris (ed.), *The Concept of development.* University of Minnesota Press, Minneapolis. Pp. 78-108.

___________. 1966. Behavioral development and comparative psychology. *Quarterly Review of Biology* **41**: 283-303.

Schoener, T. W. 1971. Theory of feeding strategies. *Annual Review of Ecology and Systematics* **2**: 369-404.

Scholander, P. F. 1956. Climatic rules. *Evolution* **10**: 339-340.

Schurmann, C. L. and van Hooff, J. A. R. A. M. 1986. Reproductive strategies of the orangutan: New data and a reconsideration of existing sociosexual models. *International Journal of Primatology* **7**(3): 265-287.

Scott, J. P. 1958. *Animal Behavior.* Natural History Library. Anchor Books, Doubleday, Garden City, New York.

Selander, R. K. 1972. Sexual selection and dimorphism in birds. In B. G. Campbell (ed.), *Sexual selection and the descent of man 1871-1971.* Aldine Press, Chicago. Pp. 180-230.

Seyfarth, R. M. 1978. Social relationships among adult male and female baboons. I. Behaviour during sexual consortship. *Behaviour* **64**: 204-226.

___________. 1987. Vocal communication and its relation to language. In B. Smuts, D. L. Cheney, R. M. Seyfarth, R. W. Wrangham and T. T. Struhsaker (eds.), *Primate societies.* University of Chicago Press, Chicago. Pp. 440-451.

Seyfarth, R. M. and Cheney, D. L. 1982. How monkeys see the world: A Review of recent research on east african vervet monkeys. In C. T. Snowdon, C. H. Brown and M. R. Peterson (eds.), *Primate communication.* Cambridge University Press, Cambridge. Pp. 239-252.

___________. and Cheney, D. L. 1986. Complexities in the study of vervet monkey grunts. In D. M. Taub and F. H. King (eds.), *Current perspectives in primate social dynamics.* Van Nostrand Reinhold Co., New York. Pp. 378-388.

___________. Cheney, D. L. and Marler, P. 1980. Vervet monkey alarm calls: Semantic communication in a free-ranging primate. *Animal Behaviour* **28**: 1070-1094.

Shafton, A. 1976. *Conditions of awareness: Subjective factors in the social adaptations of man and other primates.* Riverstone Press, Portland, Ore.

Shapiro, G. L. 1982. Sign acquisition in a home-reared/free ranging orangutan: Comparisons with other signing apes. *American Journal of Primatology* **3**: 121-129.

Sigg, H. 1986. Ranging patterns in hamadryas baboons: Evidence for a mental map. In J. G. Else and P. C. Lee (eds.), *Primate ontogeny, cognition and social behaviour.* Cambridge University Press, Cambridge. Pp. 87-90.

___________. and Stolba, A. 1981. Home range and daily march in a hamadryas baboon troop. *Folia Primatologica* **36**: 40-75.

Silk, J. B. 1978. Patterns of food sharing among mother and infant chimpanzee at Gombe National Park, Tanzania. *Folia primatologica* **29**: 129-141.

___________. 1982. Altruism among female *Macaca radiata:* Explanations and analysis of patterns of grooming and coalition formation. *Behaviour* **79**(2/4): 162-188.

Simonds, P. E. 1974. Sex differences in bonnet macaque networks and social structure. *Archives of Sexual Behavior* **31**: 151-161.

Singh, M., 1986. Primate consciousness. In N J. G. Else and Lee, P. C. (eds.), *Primate ontogeny, cognition and social behaviour.* Cambridge University Press, Cambridge.

Singh, M., and Prakash, P. 1981. *A Study of protolinguistic communication system in macaques.* Central Institute of Indian Languages, Mysore.

Skinner, B. F. 1953. *Science and human behaviour.* Macmillan, New York.

Smith, H. J., Newman, J. D. and Symmes, D. 1982. Vocal Concomitants of Affiliative Behaviour in Squirrel Monkeys. In C. T. Snowdon, C. H. Brown and M. R. Peterson (eds.), *Primate communication.* Cambridge University Press, Cambridge. Pp. 30-49.

Smuts, B. 1985. *Sex and friendship in baboons.* Aldine, New York.

Sperber, D. 1985. Anthropology and psychology: Towards an epidemiology of representations. *Man (N.S.)* **20**: 73-89.

Stelzner, J. K. 1987. *Thermal environment and behaviour of yellow baboons.* Ph.D Thesis, Cornell University.

___________. 1988. Thermal effects on movement patterns of yellow baboons. *Primates* **29**(1): 91-105.

Stelzner, J. K. and Hausfater, G. 1986. Posture, microclimate, and thermoregulation in yellow baboons. *Primates* **17**: 311-322.

Stent, G. S. 1985. Hermeneutics and the analysis of complex biological systems. In D. J. Depew, and B. H. Weber (eds.), *Evolution at a crossroads: The new biology and the new philosophy of science.* Bradford Books of the MIT Press, Mass. Pp. 209-226.

Steptoe, A., Mathews, A. and Johnston, D. 1974. The Learned control of differential temperature in the human earlobes: preliminary study. *Biological Psychology* **1**: 237-242.

Struhsaker, T. T. 1967. Auditory communication among vervet monkeys *(Cercopithecus aethiops).* In S. A. Altmann (ed.), *Social communication among primates.* University of Chicago Press, Chicago. Pp. 281-324.

___________. 1967. Social structure among vervet monkeys *(Ceropithecus aethiops). Behaviour* **29**: 83-121.

___________. 1969. Correlates of ecology and social organization among African Cercopithecines. *Folia Primatologica* **11**: 80-118.

___________. 1975. *The red colobus monkey.* University of Chicago Press, Chicago.

Strum, S. C. 1975. Primate predation: Interim report on the development of a tradition in a troop of olive baboons. *Science* **187**: 755-757.

___________. 1981. Processes and products of change: Baboon predatory behavior at Gilgil, Kenya. In R. S. O. Harding, and G. Teleki (eds.), *Omnivorous primates: Gathering and hunting in human evolution.* Columbia University Press, New York. Pp. 255-302.

___________. 1987. *Almost human: A journey into the world of baboons.* Random House, New York.

Suarez, S. and Gallup, G. G. 1981. Self-recognition in chimpanzees and orangutans, but not gorillas. *Journal of Human Evolution* **10**: 175-188.

Sugardjito, J. and Nurhuda, N. 1981. Meat-eating behaviour in wild orang-utans *(Pongo pygmaeus)*. *Primates* **22**(3): 414-416.

___________. te Boekhorst, I. J. A. and van Hooff, J. A. R. A. M. 1987. Ecological constraints on the grouping of wild orangutans *(Pongo pygmaeus)* in the Gunung Leuser National Park, Sumatra, Indonesia. *International Journal of Primatology* **8**(1): 17-41.

Sugiyama, Y. 1976. Life history of male Japaneses monkeys. In J. S. Rosenblatt, R. A. Hinde, E. Shaw and C. Beer (eds.), *Advances in the study of behaviour,* **Vol.** 7. Academic Press, New York. Pp. 255-285.

Suomi, S. J. and Harlow, H. 1972. Social rehabilitation of isolate reared monkeys. *Developmental Psychobiology* **6**(3): 487-496.

Sussman, R. W. 1979. *Primate ecology.* John Wiley, New York.

___________. 1984. The Locomotor behaviour of *Pan paniscus* in the Lomako Forest. In R. L. Susman (ed.), The *Pygmy chimpanzee: evolutionary biology and behaviour.* Plenum Press, New York. Pp. 369-393.

Suzuki, D. T., Griffiths, A. J. F., Miller, J. and Lewontin, R. C. 1989. *An introduction to genetic analysis.* Freeman and Co, New York.

Swan Productions, England. 1975. *The baboons of Gombe.* Distributed Visual Education Centre, Canada.

Swan, H. 1974. *Thermoregulation and bioenergetics.* Elsevier, New York.

Takahata, Y., Hasegawa, T. and Nishida, T. 1984. Chimpanzee predation in the Mahale Mountains from August 1979 to May 1982. *International Journal of Primatology* **5**(3): 213-233.

Taylor, H., Teas, J., Richie, T. Southwick, C. and Shrestha, P. 1978. Social interactions between adult male and infant rhesus monkeys in Nepal. *Primates* **19**(2): 343-355.

te Boekhorst, I. J. A., Schurmann, C. L. and Sugardjito, J. 1990. Residential status and seasonal movements of wild orangutans in the Gunung Leuser Reserve (Sumatra, Indonesia). *Animal Behaviour* **39**: 1098-1109.

Teleki, G. 1975. *The Predatory behavior of wild chimpanzees.* Bucknell University Press, Lewisburg.

Terborgh, J. W. 1983. *Five new world primates: A study in comparative ecology.* Princeton University Press, Princeton.

___________. 1990. Mixed flocks and polyspecific associations: costs and benefits of mixed groups of birds and monkeys. *American Journal of Primatology* **21**: 87-100.

Terborgh, J. W., and Janson, C. H. 1986. The socioecology of primate groups. *Annual Review of Ecology and Systematics* **17**: 111-35.

Tinbergen, N. 1951. *The study of instinct.* Clarendon Press, Oxford.

Tooby, J. and Devore, I. 1987. The Reconstruction of Hominid Behavioral Evolution. In W. G. Kinzey (ed.), *The evolution of human behavior: primate models.* State of New York University Press, Albany. Pp. 183-237.

Trivers, R. L. 1971. The evolution of reciprocal altruism. *Quarterly Review of Biology* **46**: 35-57.

___________. 1972. Parental investment and sexual selection. In B. G. Campbell (ed.), *Sexual selection and the descent of man 1871-1971.* Aldine Press, Chicago. Pp. 136-179.

Tsumori, A. 1967. Newly acquired behaviour and social interactions of Japanese Monkeys. In S. A. Altmann, (ed.), *Social communication among primates.* University of Chicago Press, Chicago. Pp. 207-220.

Tutin, C. E. G. and Fernandez, M. 1983. Gorillas feeding on termites in Gabon, West Africa. *Journal of Mammalogy* **64**(3): 530-531.

Tuttle, R. H. 1975. *Socioecology and psychology of primates.* Mouton, The Hague.

Van Schaik, C. 1983. Why are diurnal primates living in groups? *Behaviour* **87**: 120-144.

___________. and Van Hooff, J. A. R. 1983. On the ultimate causes of primate social systems. *Behaviour* **85**: 91-117.

Vehrencamp, S. L., and Bradbury, J. S. 1984. Mating systems and ecology. In J. R. Krebs and N. B. Davies (eds.), *Behavioural ecology: An evolutionary approach.* Blackwell, Oxford. Pp. 251-278.

Vehrencamp, S. L., Bradbury, J. W. and Gibson, R. M. 1989. The energetic costs of display in male sage grouse. *Animal Behaviour* **38**: 885-896.

Verner, J., and Willson, M. F. 1966. The influence of habitats on mating systems of North American passerine birds. *Ecology* **47**: 143-147.

Vogt, J. L. 1978. The social behavior of a marmoset *(Saguinus fuscicollis)* group III. Spatial analysis of social structure. *Folia Primatologica* **29**: 250-267.

Vrba, E. S., and Eldredge, N. 1984. Individuals, hierarchies and processes: towards a more complete evolutionary theory. *Paleobiology* **10**: 146-171.

de Waal, F. 1977. The Organization of agonistic relations within two captive groups of Java monkeys *(Macaca fascicularis). Zeitschrift fur Tierpsychologie* **44**(3): 225-282.

___________. 1982. *Chimpanzee politics: Power and sex among apes.* Jonathan Cale, London.

___________. 1986. The integration of dominance and social bondlng in primates. *Quaternary Review of Biology* **61**: 459-479.

___________. 1989. Behavioral contrasts between Bonobos and chimpanzee. In P. G. Heltne and L. Marquardt (eds.), *Understanding chimpanzees.* Chicago Academy of Sciences, special publication no. 16, Harvard University Press, Cambridge, Mass. Pp. 154-175.

___________. 1989. *Peacemaking among primates.* Harvard University Press, Cambridge.

___________. and Yoshihara, D. 1983. Reconciliation and redirected affection in rhesus monkeys. *Behaviour* **85**: 224-241.

___________. and Luttrell, L. M. 1989. Toward a comparative socioecology of the genus macaca: different dominance styles in rhesus. *American Journal of Primatology* **19: 83**-109.

Waddington, C. H. 1957. *The Strategy of the genes.* Allen and Unwin, London.

Waser, P. M. 1977. Individual recognition, intragroup cohesion and intergroup spacing: Evidence from sound playback to forest monkeys. *Behaviour* **60**: 28-74.

___________. 1982. The evolution of male loud calls among mangabeys and baboons. In C. T. Snowdon, C. H. Brown and M. R. Peterson (eds.), *Primate communication.* Cambridge University Press, Cambridge. Pp. 117-143.

Washburn, S. L. and Lancaster, J. 1968. The evolution of hunting. In R. Lee and I. Devore (eds.), *Man the hunter.* Aldine, Chicago. Pp. 293-303.

___________. and Devore, I. 1961. The social life of baboons. *Scientific American* **204**: 63-71.

___________. and Hamburg, D. A. 1968. Aggressive behavior in old world monkeys and apes. In P. Jay (ed.), *Primates: studies in adaptation and variability.* Holt, Rinehart, Winston, San Francisco. Pp. 458-478.

Wells, D. T. 1970. Large magnitude voluntary heart rate changes under conditions of high motivation and extended practise. *Psychophysiology* **6**: 636.

Wheatley, B. P. 1982. Energetics of foraging in *Macaca fascicularis* and *Pongo pygmaeus* and a selective advantage of large body size in the orangutan. *Primates* **23**(3): 348-363.

___________. 1987. The evolution of large body size in orangutans: A model for hominoid divergence. *American Journal of Primatology* **13**: 313-324.

Whitehead, J. M. 1986. Development of feeding selectivity in mantled howling monkeys, *Alouatta palliata*. In J. G. Else and P. C. Lee (eds.), *Primate ontogeny, cognition and social behaviour.* Cambridge University Press, Cambridge. Pp. 105-118.

Whiten, A. and Byrne, R. 1988a. The Machiavellian intelligence hypotheses: editorial. In R. Byrne and A. Whiten (eds.), *Machiavellian intelligence: Social expertise and the evolution of intellect in monkeys, apes and humans.* Oxford University Press, Oxford. Pp. 1-9.

___________. 1988b. Taking (Machiavellian) intelligence apart: editorial. In R. Byrne and A. Whiten (eds.), *Machiavellian intelligence: Social expertise and the evolution of intellect in monkeys, apes and humans.* Oxford University Press, Oxford. Pp. 50-65.

___________. 1988c. The manipulation of attention in primate tactical deception. In R. Byrne and A. Whiten (eds.), *Machiavellian intelligence: Social expertise and the evolution of intellect in monkeys, apes and humans.* Oxford University Press, Oxford. Pp. 211-223.

Williams, G. C. 1966. *Adaptation and natural selection.* Princeton, University Press, Princeton, N.J.

Williams, T. R. 1972. The socialization process: A theoretical perspective. In F. E. Poirier (ed.), *Primate socialization.* Random House, New York. Pp. 174-207.

___________. 1975. Introduction. In T. R. Williams (ed.), *Socialization and communication in primary groups.* Mouton, The Hague. Pp. 1-11.

Wilson, E. O. 1975. *Sociobiology: The new synthesis.* Belknap, Harvard University Press, Cambridge, Mass.

Wood, R. J. 1984. Spontaneous use of sticks by gorillas at Howletts ZooPark, England. *Int. Zoo News* **31**(3): 13-18.

Woodruff G. and Premack, D. 1979. Intentional communication in the chimpanzee: The development of deception. *Cognition* **7**: 333-362.

Wrangham, R. W. 1977. Feeding behaviour of chimpanzees in Gombe National Park, Tanzania. In T. H. Clutton-Brock (ed.), *Primate ecology: studies of feeding and ranging behaviour in lemurs, monkeys and apes.* Academic Press, London. Pp. 503-538.

__________ and Smuts, B. B. 1980. Sex differences in behavioural ecology of chimpanzees in Gombe National Park, Tanzania. *Journal of Reproduction and Fertility (Supplement)* **28**: 13-31.

Wright, P. C. 1990. Patterns of Paternal Care in Primates. *International Journal of Primatology* **11**(2): 89-102.

Wright, R. V. S. 1972. Imitative learning of flaked-tool technology - the case of an orangutan. *Mankind* **8**: 296-306.

Yamagiwa, J. 1985. Socio-sexual factors of troop fission in wild Japanese monkeys *(Macaca fuscata yakui)* on Yakushima island, Japan. *Primates* **26**: 105-120.

Yoshiba, K. 1964. Report of the preliminary survey on the orangutan in North Borneo. *Primates* **5**(1-2): 11-26.

Yoshikubo, S. 1985. Species discrimination and concept formation by rhesus monkeys *(Macaca mulatta). Primates* **26**: 285-299.

Zeller, A. C. 1980. Primate facial gestures: a study of communication. *Papers in Linguistics: International Journal of Human Communication.* **13**: 565-606.

__________. 1986. Comparison of component patterns in threatening and friendly gestures in *Macaca sylvanus* of Gibraltar. In D. M. Taub and F. A. King (eds.), *Current perspectives in primate social dynamics.* Van Nostrand Reinhold, New York. Pp. 487-504.

__________. 1987. A Comparison of component patterns in barbary macaque and japanese macaque threat gestures. Paper given at the American Society of Primatologists' Meetings. Madison, 1987.

__________. *MS.* A Film analysis of the standard threat of *Macaca sylvanus.*

Zihlman, A. L. and Tanner, N. M. 1978. Gathering and the hominid adaptation. In L. Tiger and H. M. Fowler (eds.), *Female hierarchies.* Beresford Books, Chicago. Pp. 163-194.

CONTRIBUTORS

FRANCES D. BURTON: Her field work began in Kenya, at Limuru on captive, and Thika, on wild cercopithecines. Prof. Burton conducted a longitudinal study on the macaques of Gibraltar, and more recently has been studying the hybrid macaques of New Territories, Hong Kong. Her primary focus is on social behaviour as "tradition" or "cultural" process. She teaches at the Scarborough Campus of the University of Toronto.

JOHN CHADWICK-JONES: A social psychologist at St. Mary's University, Halifax, Prof. Chadwick-Jones has recently turned his attention to the study of non-human primates. He is applying the theories and techniques of social psychology to further knowledge of the social workings of non-human primate societies.

LESLIE CHAN: Currently a lecturer at the University of Toronto's Scarborough Campus, Leslie Chan has done field work on the hybrid macaques of New Territories (Hong Kong). His interests are in evolutionary biology and on the question of speciation and how it occurs.

BIRUTE GALDIKAS: Dr. Galdikas is the acknowledged expert on orangutans, having spent the better part of her life researching them. Her efforts at rehabilitation of formerly captive orangs has not only provided a methodology by which reintroduction can be achieved, but also firmly established conservation of orang utans as a programme in Indonesia. She teaches in the Department of Archaeology at Simon Fraser University in British Columbia.

SUSAN HORNSHAW: Although originally trained in psychology, Dr. Hornshaw took her degree in anthropology. She has studied macaques, especially Lion-tails, in captivity and Barbary Apes both free-ranging and in zoos. Her interests are in methodology and the theoretical commitments that form them. Prof. Hornshaw teaches at the University of Winnipeg in Manitoba where her interests are currently on epistemological deconstruction.

JAMES PATERSON: Dr. Paterson, who has recently completed his Chairmanship with the Department of Anthropology at the University of Calgary, has done research on a variety of cercopithecoids in captivity and in the 'wild.' His original research was on baboons but more recently he has turned his attention to macaques and New World monkeys in captivity. Prof. Paterson is concerned with the effects of physiology on social behaviour.

PAUL VASEY: is completing his graduate work. His field research includes work on capuchins at Santa Rosa National Park, Costa Rica. His interests are in social intelligence as it is expressed in orang utans, and in its implications for hominization.

ANNE ZELLER: Dr. Zeller's primary focus is on social communication in non-human primates. Her analysis of the components of facial communication have alerted researchers to the fact that non-human primates have a richer "vocabulary" from which to draw and greater subtlety in the coding of statements. She has done field work on free-ranging and wild macaques in Gibraltar and Indonesia and on captive ones as well. She teaches at the University of Waterloo.

INDEX

S

T

V

INDEX OF AUTHORS

N

O

P

Q

R

S

Old Adult Female "Wilma" Macaca sylvanus

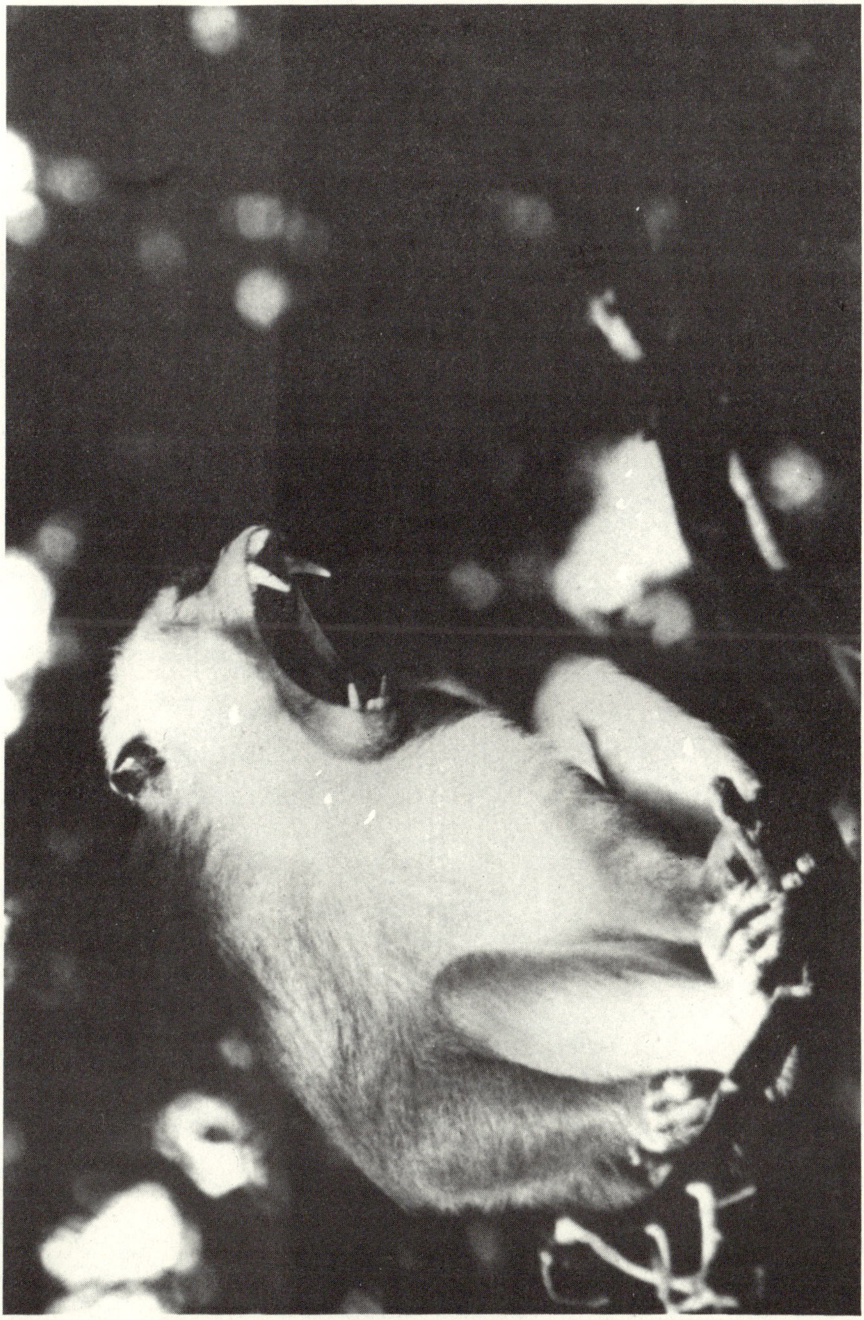

M. fascicularis

Adult Male orangutan, Pongo pygmaeus.

Adult male & 2 juveniles. Macaca silenus

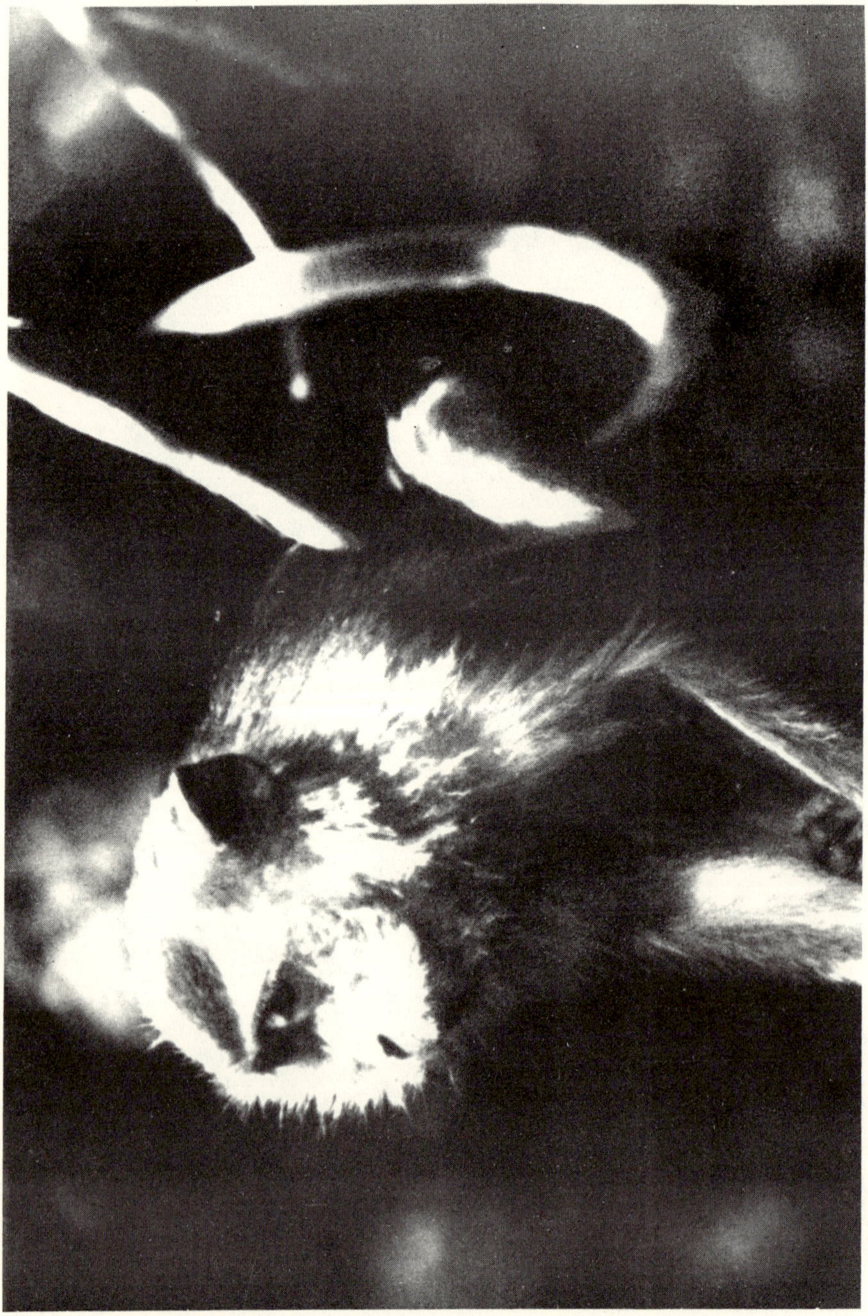

Macaca fascicularis - Subadult

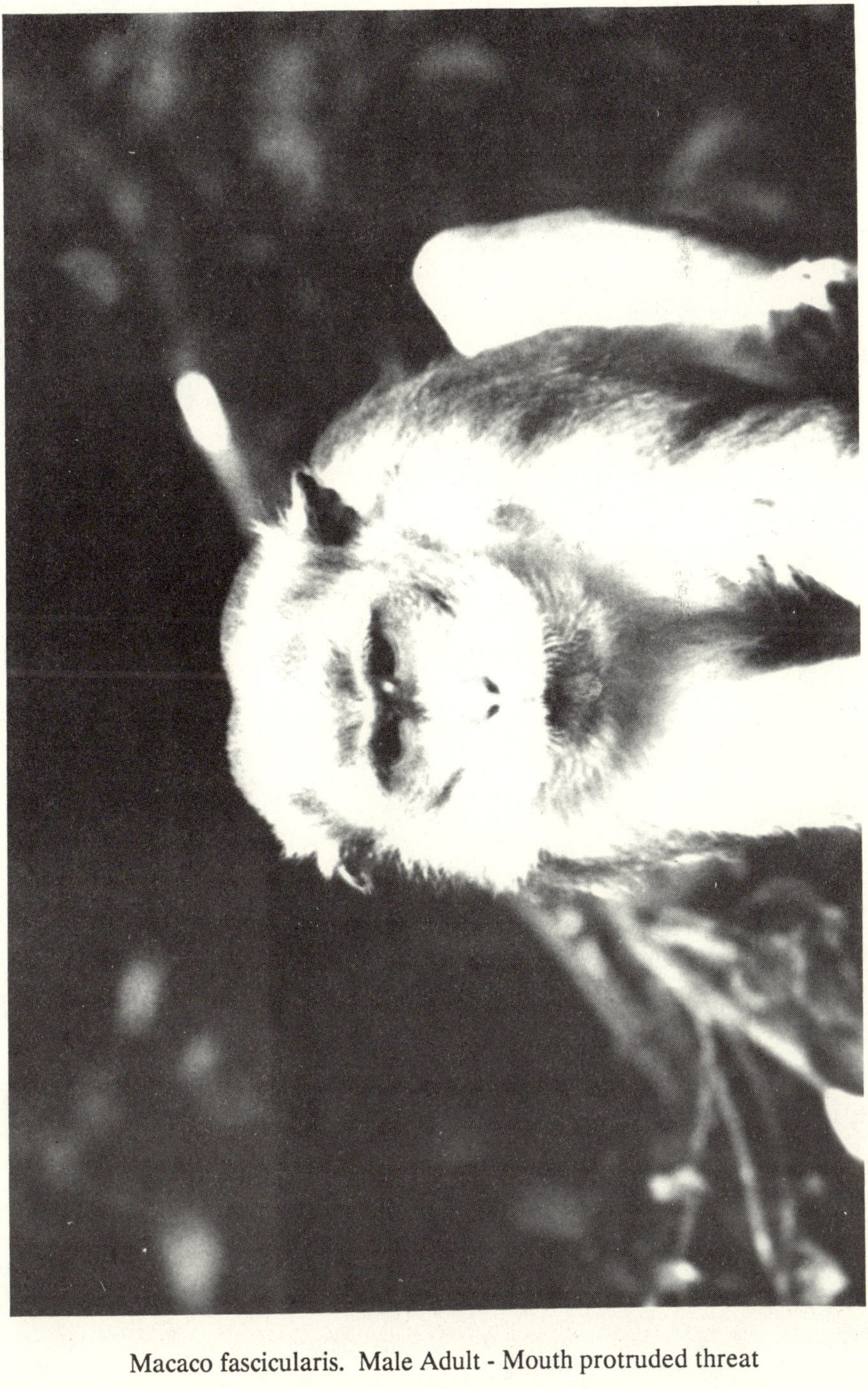

Macaco fascicularis. Male Adult - Mouth protruded threat